LOSING
GROUND

LOSING GROUND

Environmental Stress and

World Food Prospects

ERIK P. ECKHOLM

WORLDWATCH INSTITUTE, WITH THE SUPPORT AND
COOPERATION OF THE UNITED NATIONS ENVIRONMENT PROGRAM

W · W · NORTON & COMPANY · INC ·
NEW YORK

THE TEXT *of this book is set in Avanta, an RCA Videocomp typeface.
Composition is by Haddon Craftsmen.*

Library of Congress Cataloging in Publication Data
Eckholm, Erik P

    Losing ground.
    Bibliography: p.
    Includes index.
    1. Agricultural ecology.   2. Man—Influence on
nature.   3. Food supply.   4. Soil erosion.   I. Title
S601.E24   333.7'5   75-41397

ISBN 0 393 06410 7 cloth edition
ISBN 0 393 09167 8 paper edition

2 3 4 5 6 7 8 9 0

To my father, Wendell Eckholm

Contents

Foreword

BY MAURICE F. STRONG

Executive Director, United Nations Environment Program

DURING THE COURSE of preparations for the United Nations Conference on the Human Environment held in Stockholm in June, 1972, national governments and the scientific community were asked to help identify and evaluate possible environmental issues requiring attention and action by the international community. Assessment of the national reports and scientific studies prepared for the conference made it clear that the loss of productive soil through erosion, salination, desertification, and other consequences of ecologically unsound land use were seen as one of the principal environmental problems confronting a significant number of countries, particularly in the developing world.

In many of these areas the problem was not a new one. And it was generally seen as primarily a local or national concern. What was new about the evidence assembled for the Stockholm Conference was the very scale and magnitude of the problem and the degree to which the pressures on productive soil are being intensified in the areas in which the need for increased food production is growing most rapidly.

This new publication surveys and analyzes on a systematic basis the various ways in which our delicately balanced food systems are being ecologically undermined. It provides concrete examples of the serious ecological deterioration that is taking place through deforestation, overgrazing, soil erosion and abandonment, desertification, the silting of irrigation systems and reservoirs, and the changes in the frequency and severity of flooding. In all of these, the increasing intensity of humanity's pressure on the land and the continuation of careless and short-sighted land use practices compound the effects of such natural phenomena as droughts and floods, often turning the temporary problem they create to large-scale disaster.

"Losing ground" on the scale pointed up by this book poses a serious threat to the world's capacity to feed itself in the future. Thus I welcome this as a timely and valuable contribution to public awareness and understanding of what we must now regard as one of the principal ecological problems facing mankind. The book documents convincingly the need for priority attention to this issue and points up the kinds of actions required if we are to win the battle against world hunger.

The problem of hunger manifests itself not only in terms of starvation and death but also in the mental deprivation that results from protein and vitamin deficiency. The world community, through a series of UN conferences and programs, is seriously and genuinely attempting to respond to the world food crisis.

While the views presented in the book are those of the author, many of the issues dealt with in it figure prominently in the program of the United Nations Environment Program, for which the task of developing and supporting international measures to deal with these issues is amongst the highest priorities. UNEP's support of the preparation and publishing of this book by the Worldwatch Institute is one of these measures.

Mr. Eckholm's cogent and well-reasoned analysis of the nature and dimensions of this threat to future world food supplies makes a compelling case for the urgency of international cooperation to deal with it. And it dramatizes the direct and intimate relevance of sound environmental management to one of the central concerns confronting the human community—the need to manage the precious resources of our "Only One Earth" so as to assure provision of the basic needs of all its inhabitants.

Losing Ground then goes to the heart of the issue of human survival and it makes it clear that the issue will be decided by what people and their governments do—or fail to do—at this point in our history.

An Introductory Note

LOSING GROUND is an effort to draw attention to a set of negative ecological trends whose consequences demand far more attention than they have received to date. Ideally, a book on the ecological undermining of food-production systems would include detailed national statistics on various facets of the problem, such as overgrazing, desert encroachment, deforestation, soil erosion, flood trends, and the silting of irrigation reservoirs. Unfortunately, if understandably, such comprehensive data are not available. Governments seldom gather systematic data on negative ecological developments; the problems achieve prominence only when large-scale disaster strikes, as it recently has in some areas of sub-Saharan Africa.

Writing this book thus required a far-flung, ambitious research effort to glean relevant information from individuals and writings in a broad spectrum of disciplines. Drawing conclusions from the available data might be compared to interpreting a puzzle of which enough pieces have been assembled to provide a general impression of the overall picture. And the picture that emerges is an unsettling one that calls for a strong response by governments everywhere.

Worldwatch Institute co-sponsored *Losing Ground* in keeping with its goal of anticipating and drawing attention to global threats to human well-being. Governed by an international board of directors, the non-profit institute is located in Washington, D.C. Its program of research and public education is sponsored by founda-

tions, United Nations and governmental agencies, and private individuals. Worldwatch Institute is grateful to the United Nations
Environment Program for jointly sponsoring the research and writing
of *Losing Ground.*

Lester R. Brown,
President, Worldwatch Institute

1776 Massachusetts Avenue, N.W.
Washington, D.C. 20036

Acknowledgments

W<small>HILE</small> effusive appreciations preface many books, I feel safe in claiming that few authors have enjoyed such extensive and valuable support as I have in the conceptualization, writing, and publication of *Losing Ground.* Worldwatch Institute President Lester Brown helped originally to formulate the issues, and his constant encouragement and ready store of fresh ideas strongly influenced the book's final scope and flavor. Kathleen Newland participated from the project's inception in planning our investigation and in analyzing many of the topics covered; her wide-ranging research effort is reflected in every chapter. Kathleen Courrier's graceful editing made my prose clearer and more economical.

Blondeen Duhaney's indispensable administrative and secretarial help greatly improved the efficiency of both our research efforts and the rapid writing and rewriting of the manuscript. Other Worldwatch colleagues, Marion Frayman, Trudy Todd, and Joan Zwierchowski, also provided much-appreciated secretarial assistance.

John Tidd helped with portions of the research, particularly those concerning Soviet agriculture. Along with Anne Harrington and Rosemarie Philips, he also assisted with translations of foreign language sources. Boyd Compton's investigations in Southeast Asia were vital to making sense out of the contradictory published accounts of ecological trends in that region. I am indebted to the many individuals in Asia and West Africa whose hospitality and advice made my weeks of travel such a worthwhile addition to the research effort.

Some of the basic ecological challenges noted in this book have not been discussed extensively in either the academic literature or available government documents. Accordingly, we contacted more than a hundred ecologists, foresters, agricultural economists, academics, and national and international civil servants throughout the world, seeking their advice and assistance. The response was both invaluable and encouraging: scores generously shared their personal observations, published and unpublished papers, and suggestions about other information sources. I hope that readers of this book with pertinent ideas and experiences will perpetuate this process of informal intellectual interaction.

The United Nations Environment Program, under the leadership of Maurice Strong, joined Worldwatch in sponsoring this book and the substantial research effort it embodies. Since its establishment by the 1972 Stockholm Conference on the Human Environment, UNEP has increasingly helped alert the world to the urgency of environmentally sound development, and I am grateful for its support and cooperation.

Mr. Strong, Mostafa Tolba, and other members of the UNEP staff provided a useful review of the book's first draft. Others I want to thank for preliminary readings of the manuscript include Bruce Stokes, Mary Elmendorf, Peter Freeman, Lincoln Gordon, Dexter Hinckley, Frances Irwin, Vladimir Kollantai, Charles Pearson, David Pimentel, Andrew Rice, and B. B. Vohra. Selected chapters were reviewed by Charles Bailey, Barry Bishop, Reid Bryson, Paul Ehrlich, D. J. Greenland, Denis Hayes, M. Kassas, Patricia McGrath, Robert Potter, John Sanders, Robert Stein, Ralph Townley, and Robert Winters. Each of the reviewers contributed to the manuscript's improvement; any remaining errors are, of course, my responsibility alone.

ERIK P. ECKHOLM

LOSING
GROUND

1. The Undermining of Food-Production Systems

COAXING enough food from the earth has traditionally been guided by a certain simple logic: plow more land, intensify labor, refine techniques, and the supply of food will grow commensurately. But this has been the logic of humans, not of nature, and today's newspaper headlines tell with increasing frequency a different, more puzzling story. Millions of individuals, and sometimes whole countries, are learning the hard way that more work doesn't necessarily mean more food—that it may mean fatally less.

A Somali nomad builds his herd to record size, but the grassland is overgrazed, his cattle grow thin, and sand dunes bury pastures. A farmer in northern Pakistan clears trees from a mountain slope to plant his wheat; soon after, fields downstream are devastated by severe floods. In Indonesia, a peasant burns away luxurious hillside vegetation to plant his seeds; below, rice production drops as soil washed down the mountain chokes irrigation canals.

Over the course of ten thousand years humans have successfully learned to exploit ecological systems for sustenance. Nature has been shaped and contorted to channel a higher than usual share of its energies into manufacturing the few products humans find useful. But while ecological systems are supple, they can snap viciously when bent too far. The land's ability to serve human ends can be markedly, and sometimes permanently, sapped.

The international discussion of environmental quality has, like that of many other topics, been largely pre-empted by the rich industrial world. The term "environmental crisis" joined the lexicon of journalism and politics only within the last decade, in response to the visible spread of acrid air and poisoned waters. Even within the field of agriculture, concern for ecological damage usually focuses on the polluting impact of misused chemicals.

These problems are pressing enough, and deserve all the attention they have received and more. Yet in the world war to save a habitable environment, even the battles to purify the noxious clouds over Tokyo and São Paulo, and to rescue life in Lake Erie, are but skirmishes compared to the uncontested routs being suffered in the hills of Nepal and Java, and on the rangelands of Chad and northwest India. A far deadlier annual toll, and perhaps an even greater threat to future human welfare, than that of the pollution of our air and water is that exacted by the undermining of the productivity of the land itself through accelerated soil erosion, creeping deserts, increased flooding, and declining soil fertility. Humans are—out of desperation, ignorance, shortsightedness, or greed—destroying the basis of their own livelihood as they violate the limits of natural systems.

Not surprisingly, the principal victims of these trends are the world's poor, who, in their quest for food and fuel, are often forced by circumstances and institutions beyond their control to serve as the agents of their own undoing. Though poverty is often associated with a pristine environment, and affluence with despoliation, in some important ways the poor are damaging the environment even more than the rich.

The littered ruins and barren landscapes left by dozens of former civilizations remind us that humans have been undercutting their own welfare for thousands of years. What is new today is the awesome scale and dizzying speed with which environmental destruction is occurring in many parts of the world. The basic arithmetic of world population growth reveals that the relationship between human beings and the environment is now entering an historically unique age of widespread danger. Whatever the root causes of suicidal land treatment and rapid population growth—and the causes of both are numerous and complex—in nearly every instance the rise in human numbers is the immediate catalyst of deteriorating food-production systems. The number of humans reached one billion about 1830, two or three million years after our emergence as a distinct species. The second billion was added in one hundred years, and the third billion in thirty years. One day in late 1975, just fifteen years later, world population reached four billion. At the present rate of growth, the fifth billion will come in thirteen years and the sixth in ten years after that.

Seldom does the imagination translate these inconceivable abstractions into the events on the ground that give them meaning: farmers forced onto mountain slopes so steep that crops and topsoil wash away within a year; peasants making charcoal out of forests that are essential for restraining flood waters and soil erosion; drought-prone pastures plowed up for grain despite the high odds that a lifeless dust bowl will ensue. In some respects, these are Malthusian phenomena with a twist. Exponentially growing populations not only confront a fixed supply of arable land, but sometimes they also cause its quality to diminish. However, a second addendum to Malthus' gloomy formulation is also crucial. Today the human species has the knowledge of past mistakes, and the analytical and technical skills, to halt destructive trends and to provide an adequate diet for all using lands well-suited for agriculture. The mounting destruction of the earth's life-supporting capacity is not the product of a preordained, inescapable human predicament, nor does a reversal of the downward slide depend upon magical scientific breakthroughs. Political and economic factors, not scientific research, will determine whether or not the wisdom accumulating in our libraries will be put into practice.

This book is not the first effort to spotlight self-defeating efforts to expand the supply of food, and it most certainly will not be the last. Earlier in this century a number of analysts, concerned primarily with the threat of soil erosion, catalogued the ecological calamities impending if humans did not change their ways.[1] Erosion was sometimes painted as the greatest single threat to the future of human civilization. In the 1930s, as the Great Plains heartland of North America wasted away into the Dust Bowl—one of the century's more dramatic environmental debacles—previous warnings achieved an air of prophecy.

Then, as the Great Plains were recovering from their human-caused infirmity, World War II created environmental disasters of a different type and scale, and was followed by an era of economic expansion such as the world had never before seen. World agriculture, too, entered an era of remarkable gains, with global grain output far outpacing population growth since the late forties. Headlong economic growth in the richer countries, and in small pockets of the poorer countries, has created new environmental challenges as the refuse of industrialism piles up, but earlier predictions that land degradation would be humanity's downfall seem to have been disproved by history.

Or have they? The continuing growth in world food output and the remarkable climb of the gross national product in most countries, rich and poor, over the last two decades mask some basic facts that add up to a different story. National income averages conceal the billion or more people locked in cycles of abject poverty, misery, and exploitation, many of whom live in worse conditions than their parents did. World, and even national, food output totals conceal the stagnant or deteriorating productivity of huge numbers of farmers in the poorer regions and countries. Such figures veil the half-billion people suffering chronic malnutrition in the best of years, and the hundreds of millions who join their lot when food prices soar, as they have in the mid-seventies. The statistics of progress ignore the swelling urban shantytowns filled with refugees from untenable rural situations. In short, the aggregate figures for growth, both of agriculture and economies, disregard the casualties and the cast-offs of the global development process.

The wretched lot of one-fourth or more of the world's people has not, of course, gone unnoticed. Through both national and international channels, many billions of dollars and the talents of thousands of "experts" are earmarked each year for the agricultural progress of the poor. Broadly based rural development through agriculture, it is increasingly recognized, can ameliorate simultaneously many of the world's most acute social problems. Not only will more food be grown under the auspices of such programs, but it will be grown by those who most need more nourishment and income. By providing greater income and employment in rural areas, properly designed agricultural development can help stem the flow of migrants to the cities. And with improved social conditions, the rural poor may well follow the historical path toward smaller families. Yet somehow, in many regions of many countries, things don't seem to be working. All the money, all the research, all the experts have done little for those on the bottom. What has gone wrong?

Looking at this dark side of the development record, analysts find many culprits. Depending on personal prejudices, the economist may see a failure to generate adequate capital for raising productivity; imperfect markets for labor, capital, and technology that impede efficiency; or systems of trade and investment heavily weighted against the well-being of the poor. The sociologist may see tradition-fixed cultures incapable of assimilating the requirements of "modernization," or socio-economic structures that compel the poor to live recklessly. The political scientist may stress the absence of the administrative capacity to implement social change, or outline the power relationships that prevent the poor from taking control of their own destiny.

All these explanations contain truths, but an additional perspective that in many ways subsumes them all, and is almost always overlooked, is an ecological view that blends the study of human beings' place in nature with that of their place in society. A common factor linking virtually every region of acute poverty, virtually every rural homeland abandoned by destitute urban squatters, is a deteriorating natural environment. Ecological degradation is to a great extent the *result* of the economic, social, and political inadequacies noted above; it is also, and with growing force, a principal *cause* of poverty. If the environmental balance is disturbed, and the ecosystem's capacity to meet human needs is crippled, the plight of those living directly off the land worsens, and recovery and development efforts—whatever their political and financial backing—become all the more difficult. The soft underbelly of global rural development efforts, environmental deterioration is an often neglected factor that severely undermines their effectiveness.

The glaring disregard of the ecological requisites of progress is at least partially attributable to the rigid compartmentalization of professions, both in the academic world and in governmental agencies. When reading the analyses of economists, foresters, engineers, agronomists, and ecologists, it is sometimes hard to believe that all are attempting to describe the same country. The actions of experts frequently show the same lack of mutual understanding and integration. Engineers build one dam after another, paying only modest heed to the farming practices and deforestation upstream that will, by influencing river silt loads, determine the dams' lifespan. Agricultural economists project regional food production far into the future using elaborate, computerized models, but without taking into account the deteriorating soil quality or the mounting frequency of floods that will undercut it. Water resource specialists sink wells on the desert fringes with no arrangement to control nearby herd sizes, thus ensuring overgrazing and the creation of new tracts of desert. Foresters who must plant and protect trees among the livestock and firewood gatherers of the rural peasantry receive excellent training in botany and silviculture, but none in rural sociology; their saplings are destroyed by cattle, goats, and firewood seekers within weeks after planting.

A failure to place agriculture in its ecological context has been apparent at even the highest levels of global policy-making. Nowhere were forests so much as mentioned in the dozens of resolutions directed to eliminating hunger passed by the Rome World Food Conference of November, 1974, despite the accelerating deforestation of Africa, Asia, and Latin America and its myriad effects on food production prospects. The editors of the United Nations magazine *Ceres* were not speculating idly when they wrote in early 1975: "It is no coincidence that the forests of all the countries with major crop failures in recent years due to droughts or floods—Bangladesh, Ethiopia, India, Pakistan, and the Sahel countries—had been razed to the ground."[2]

Even without the impediment of professional or disciplinary blinders, recognizing and controlling the causes of environmental deterioration present special problems. Precise figures on ecological trends and their impact on agricultural systems are scarce. This does not make the trends any less real or menacing; it does suggest the exceptional difficulty of isolating and measuring these factors.

The symptoms of ecological stress are often confused with their causes, in part because the more visible environmental disasters are usually precipitated by a harsh turn of nature, such as prolonged drought or excessive rainfall. When topsoil clouds the air in the Great Plains or the African Sahel, or when record floods rise in India, it is tempting to place full blame on nature. It is easy to ignore the human role in making a region vulnerable to damage far more drastic than inclement weather alone would cause. Under some conditions the stresses mount almost invisibly, gradually building force until the system suddenly explodes with an unexpected fury. This has happened before, and will happen again. The question to be answered is not whether but where future ecological debacles will detonate—and with what casualties.

World fisheries present an instructive, readily measured example of what happens when too much is demanded of a food-producing ecosystem. In this case it is more often the rich than the poor who have neglected ecological reality and reaped the penalties. While the differences between farming the land and extracting food from the oceans are obvious, declining catches in numerous regional fisheries demonstrate clearly that greater effort and investment can bring not just diminishing returns but, as fish stocks are depleted by overfishing, *negative* returns. And in the early seventies, the sum of these local pressures produced a three-year sustained drop in the total *global* fish catch. We must now ask how many localized land regions have already experienced similar absolute declines in food output due to environmental degradation—and how many entire countries may be following suit, as some in central Africa apparently have already.

Rapid population growth, miserable social conditions, and environmental deterioration form the ultimate vicious circle. Improved living conditions and economic security encourage smaller families, yet steeply climbing populations undermine the effort to provide adequate nutrition, health care, education, and housing, and drain funds and energies from productive investment. New technologies can often increase the ability of even fragile ecosystems to produce food or other goods for humans, but burgeoning populations bound by poverty and traditional techniques can drastically impair the land's life-support capacity.

The sterile debate among those who advocate attacking this conundrum mainly from one side or another grows more shrill with each passing year. Simply inundate the poor with birth control devices, some say, and development prospects will soon improve. Concentrate on social reforms and economic progress, others say, and the population problem will take care of itself. Many forget that these issues form a circle, not a square, and thus have no distinct sides. The only alternative at this stage of human history is to simultaneously meet this quandary at every point along its circumference, in an all-out effort to turn the negative chain reactions into positive ones.

Billions of human beings are still denied access to family planning services, though surveys of parents in even the poorest countries often reveal a gap between desired and actual family size. A redistribution of power, land, and social services can improve the prospects of the world's dispossessed and also pull down birth rates. Yet reform and development efforts will not achieve their aims if they are not also suffused with an ecological ethic that recognizes the conjugal bond between humankind and the natural world from which there can be no divorce. Environmental deterioration requires direct attention in its own right; at the same time, the balance of nature will not be preserved if the roots of poverty, whatever they may be, are not eradicated.

2. A History of Deforestation

Though concrete and plastic blind some people to their dependence upon trees, the debt of all humans to the forest remains almost filial. Paper, taken for granted until its supply runs short, is nearly as crucial to modern societies as food or fuel, for it is the feed-stock of government, business, education, and most interpersonal communication. Wood remains the principal cooking fuel in Africa, Asia, and Latin America, as well as the sole source of warmth in cold mountain regions of these continents. Timber is still a basic raw material for construction everywhere, and cellulose, which forms the bulk of wood, assumes such useful guises as cellophane, rayon, and plastic. Forest recreation areas and natural preserves provide needed psychic shelter as societies urbanize.

For economic reasons alone, countries with little forest area are saddled with a great burden. The price of wood for furniture or construction is sometimes far higher in wood-short poor countries than it is in the United States or Europe, where personal incomes are many times higher. Meeting basic needs for firewood or charcoal drains both the national economy and family incomes when lengthy wood-gathering journeys consume the labor of the poor. Paper is always a costly import item when domestic production cannot keep up with demand; when global investment in this capital-intensive industry lags behind the need, as it has in recent years, acute shortages develop in poorer importing countries and sorely hamper communications. Where wood is exceptionally scarce, even the construction of basic infrastructure like railroad tracks and power lines is constrained.

The crucial ecological roles of forests are even less visible to most than their social and economic services. Forests influence the wind, temperature, humidity, soil, and water in ways often discovered only after the trees are cut, and these functions—usually beneficial to people—are sabotaged. Forests assist in the essential global recycling of water, oxygen, carbon, and nitrogen—and without any expenditure of irreplaceable fossil fuels. Rainwater falling on tree-covered land tends to soak into the ground rather than to rush off; erosion and flooding are thus reduced, and more water is likely to seep into underground pools and springs.

The history of the earth's forests in the billions of years preceding the emergence of *Homo sapiens* is shrouded in the leisurely processes of biological evolution, climatic change, and the rise, fall, and drift of land masses—all of which continue today. The land now called the Sahara Desert, for example, was probably located near the South Pole 450 million years ago and has slowly passed through every major climatic zone. Now drifting northward at a rate of one to two centimeters per year, the Sahara will shift one degree in latitude over the next five to ten million years, and will change climate and vegetation accordingly.[1]

The appearance of human-like creatures in the last few million years marked the beginning of a new biological age, for these animals were the first to willfully alter the environment for self-gratification. The human ability to do so grew almost imperceptibly for most of our species' tenure on earth, but eventually, in the last 1 percent of that time, it mushroomed until man came to rival the weather as a shaper of the vegetative environment—and in some regions to surpass it.

Though it understandably remains unwritten, a complete history of the earth's forests in the period since man evolved would be a fascinating analogue to the history of human civilization. Man's treatment of forests is deeply enmeshed in the history of technologies, empires, and ideas, as the fate of one well-known and well-studied stand of trees, the cedars of Lebanon, teaches.

Mount Lebanon, the backbone of the country of the same name, was once carpeted with a rich stand of stately cedars whose utility for human ends, in a part of the world where few tall, strong trees existed, became legendary throughout the Old World, and was finally their undoing. Where the cedars did not grow, the slopes of this range were filled with pine, fir, juniper, and oak. Today only scattered remnants of these once extensive forests endure. The cedar is limited to twelve small, widely separated groves that were protected over the millenia by either their inaccessibility or their proximity to chapels. Most of Mount Lebanon is now, in the words of Marvin W. Mikesell, who has chronicled its biological abasement, "as barren as the mountains of the Sahara."[2] Scrubby vegetation stubbornly survives amid rocky patches off which the soil has washed. Five thousand years of service to civilization has left the Lebanese highlands a permanently degraded vestige of their former glory.

The felling of the cedars began as early as 3000 B.C. when a Semitic desert tribe, later known as the Phoenicians, occupied the eastern shore of the Mediterranean. In a mountainous land of limited agricultural potential, the Phoenicians took to crafts and commerce, and their trading ships sailed as far as Britain and West Africa. Their glass, dyes, textiles, pottery, and metals helped nurture the prosperity of port cities like Tyre and Sidon, and before long these became trading hubs for the goods of most of the Mediterranean region and southwest Asia. Of all Phoenicia's products, perhaps the most highly prized were its cedars. Valued for strength and size, used for shipbuilding and palatial architecture, these trees were coveted by dozens of successive Old World civilizations, most of which faced chronic shortages of high-quality wood.

The Pharaohs of ancient Egypt were the first documented importers of Mount Lebanon cedars. Records of the Pharaoh Snefru from about 2600 B.C. acknowledge the arrival of forty ships filled with cedar, purchased to build a ship and palace doors. For the next two thousand years timber found its way to Egypt, either as barter or, whenever Egypt was able to maintain hegemony over Phoenicia, as tribute.

The cedars of Lebanon were also prized in Mesopotamia, the other cradle of civilization in the Middle East. A vast green mountain with tall cedars is alluded to in the *Epic of Gilgamesh*, the famous Mesopotamian myth of the third millenium B.C., but the first proof of a timber trade comes about 1100 B.C., when the conquering Assyrians demanded a steady tribute of cedar wood for the construction of temples and palaces. Successive Mesopotamians kept up the practice, often hauling logs to the Euphrates and floating them home. Nebuchadnezzar, the Babylonian King of the sixth century B.C., even left inscriptions on Mount Lebanon boasting of his achievement in securing such fine wood from afar for his kingdom.

The story of the exhaustion of this valuable resource by successive civilizations has many chapters. Biblical references to Palestinian cedar imports tell of Solomon, King of Israel, sending to Hiram, King of Tyre, the following message about 950 B.C.: "As you dealt with David my father and sent him cedar to build himself a house to dwell in, so deal with me." Thousands of laborers traveled to Lebanon in relays to obtain timber for Solomon's famous temple in Jerusalem. In return Solomon sent 125,000 bushels of wheat and a million gallons of olive oil.[3] Six hundred years later, Alexander of Macedon probably used Lebanese cedars in his Euphrates fleet, and Antigonus, one of his successors, reportedly sent eight thousand laborers and a thousand pairs of draft animals to Mount Lebanon to obtain lumber for his fleet.

After controlling Lebanon for a century and a half, Rome finally attempted to protect the remaining stands of cedar; the Emperor Hadrian (117–138 A.D.) placed nearly one hundred rock inscriptions on the northern half of Mount Lebanon, designating the surviving forests as an imperial domain. The pace of exploitation probably slowed for several centuries thereafter, though likely due as much to the diminished supply as to any restraint on the part of the region's rulers.

New threats to the forests of Mount Lebanon emerged when settlers, particularly persecuted religious sects, and their goats began moving onto its slopes in the seventh century. By this time much of the range was already denuded, and while various conquerors ever since have gathered what timber they could, perhaps the prime agents of forest degradation over the last thousand years have been firewood traders, who to this day sell charcoal from Mount Lebanon in Damascus, Tripoli, and other cities of the region. The last large-scale siege on the forests of Lebanon occurred during World War II, when British forces cut firs and oaks from the northwestern side of the mountain for use as ties on the railroad they built linking Tripoli and Haifa.

The cedars of Lebanon did not come under serious assault until the human societies surrounding them reached a rather advanced state; productive agricultural techniques and complex social organization were a prerequisite of cities, ship- and temple-building, and imperial adventures. The pace of forest destruction quickened on every continent with the emergence of agriculture and organized societies, but human influence on the world's forests has a much longer history, reaching back through hundreds of thousands of years before the invention of agriculture some ten thousand years ago. Well before the age of farms, axes, and armies, humans were altering the face of the earth, and their chief agent for doing so was fire.

While occasional fires are a natural, often beneficial factor in ecosystems, prehistoric man accidentally as well as deliberately influenced the vegetational pattern on every continent with fire. Long before he learned to manufacture flames, man almost certainly learned to tend and transport fires started by lightning or volcanos. Long after crude fire-starting techniques like the fire drill were developed, the preservation of fire probably still remained a semi-sacred chore, and even traveling groups most likely left their campfires smoldering on the chance that a spark might be left on their return journey. The logical conclusion is that, at least in drier regions and seasons, accidental forest or brush fires must have been frequent in prehistoric times, and could well have influenced the vegetation over fairly large areas.[4] Even modern man, with no incentive to do so, leaves campfires unextinguished often enough to cause massive forest damage.

These relatively haphazard changes in the extent and nature of forests were soon dwarfed by the calculated use of fire as a tool by hunters. Given the sparsity of early populations, the consumption of wood for cooking game was probably insignificant compared to the destruction of woodlands in the hunt. Fire, a time-honored weapon for driving animals out of the jungle or woods into the open or into an ambush for the kill, is a hunting technique still used today in some places, since woods and grasslands opened up and maintained by burning generally offer better and safer hunting than dense forests.

Once grazing animals were domesticated, the firing of wooded areas and grasslands probably accelerated; fires clear annoying underbrush, and fresh new grass usually springs to life after a well-timed burn. When such fires are set too often, or go out of control, the consequences are less benign. Many ecologists currently see the reckless use of fire by herders as a major cause of desert expansion and soil erosion in semi-arid regions.

In wooded areas, fire has traditionally been the chief means of clearing land for farming. This was historically true in Europe and North America, and remains so in much of Asia, Africa, and Latin America. Especially widespread today in the humid tropical forests, controlled fires are used by farmers to open up plots that are farmed for a few years and then abandoned when soil fertility dissipates.

In North America the Plains Indians "very considerably extended the prairies," notes F. Fraser Darling, by deliberately burning forests in order to expand the range of the buffalo. Lewis and Clark, on their famed expedition through the northern Rocky Mountains in 1803, found extensive prairies in areas where trees can thrive. Once the burnings stopped, forests crept back into these regions. Magellan, passing through the strait that bears his name on his historic cruise of 1520, was so impressed by the number of fires he saw that he dubbed the surrounding region Tierra del Fuego, or "Land of Fire."[5] The name would have fit most of South America, for fire was then a prevalent tool of hunters and pastoralists.

Fire has altered the face of the land most profoundly and extensively in Africa. There at least a third of the continent is covered with savanna grasslands, many of which would be forested had they been left unmolested by man. Centuries of burnings by hunters, herders, and tillers have created an enormous new ecological zone between the remaining forests and the drier natural grasslands. By 1948, one author estimated, the forests of tropical Africa had been reduced to just a third of their former domain—and the rate of deforestation has climbed in the last quarter-century.[6]

While the slow and localized destruction of forests by hunters and farmers continued throughout most of Africa, the Americas, and Southeast Asia, forest destruction began in earnest in much of the Old World with the emergence of large organized societies. The history of Mt. Lebanon and the eastern Mediterranean has already been recounted; as that story unfolded, deforestation was also accelerating throughout the Mediterranean region and in parts of Asia —particularly in China.

Most of China, which covers one-twelfth of the earth's land surface, is mountainous, and perhaps half its area was at one time forested. Long the world's most populous country, China has over the course of more than fifty-five hundred years developed gargantuan food and fuel needs. As a result, China is now struggling to overcome the dire ecological consequences of millenia of forest destruction. By the mid-twentieth century, said the Chinese minister of forestry in 1956, the country had "the greatest number of barren hills in the world."[7] Throughout much of the country only temple groves remained as a reminder of the original tree-covered landscape.

Even before Phoenicia was settled, people were moving into the fertile, heavily wooded basin of the Yellow River, or Hwang Ho, in North China. The pressing need for ever more farmland over the centuries resulted in the eventual clearing of most of the North China plain. The trend was partially arrested during the 872 years of the Chou Dynasty (1127–255 B.C.); this golden age produced what must have been the world's first "mountain and forest service," and careful attention was given forest conservation needs.[8] But widespread forest destruction again became the norm for twenty-two centuries following the Chou Dynasty's fall.

By the fourth and fifth centuries A.D., numbers of people were spreading southward to the basins of the Yangtze and Pearl Rivers, cutting forests as they went. The development of productive wet-rice cultivation sustained a large, dense population, which in turn intensified the pressures on woodlands. Industrialization and imperialism have made further inroads into China's forest lands. The Japanese cut deeply into Manchuria's rich forests during the decades they dominated, and then occupied this region in the early twentieth century. By mid-century, perhaps only 8 percent of China was forested.

Heavy erosion, accentuated by denudation of the mountains and river basins, has given some of China's rivers the highest sediment loads in the world. The beds of the Yellow and other rivers have consequently risen far above the surrounding countryside, and their waters are contained only by extensive dikes. Catastrophic floods are a constant threat that all too frequently materializes, which is why the Yellow River has long been known as "China's Sorrow." Soil erosion, siltation, and wood scarcity—all greatly exacerbated by the

loss of forests—collectively constituted one of the major challenges
facing the new Chinese government when it took power in 1949, and
forest restoration has been a major preoccupation of the government
ever since. China's population, which was about one hundred million
in the early sixteenth century, had surpassed five hundred million in
1949; today it is probably more than eight hundred million, which
makes forest regeneration an extraordinarily difficult undertaking,
and China's apparent progress in this respect an impressive achievement.

The Mediterranean climate of southern Europe and North Africa
is not conducive to dense forests, but the bare hills that characterize
these regions today provide little hint of the extensive woodlands that
once existed. By the end of the classical age, deforestation in the
lowlands around the Mediterranean was acute.[9] The clearance of
farmland, grazing herds, and wood gathering for fuel and construction all contributed to this condition. The region's dry climate and
nimble goats discouraged natural forest regeneration, even in centuries when the pressures of civilization slackened.

In the *Critias,* Plato lamented the consequences of denudation
in Attica, which were already severe by the fourth century B.C. Attica
once had much forest land, he wrote, but now "our land, compared
with what it was, is like the skeleton of a body wasted by disease. The
plump soft parts have vanished, and all that remains is the bare
carcass." Valuable springs had dried up as rainwater ran off the bare
earth before it could soak into the ground. Early a population center,
Attica reached this sorry state several centuries before most of Greece
and its neighboring lands. The eventual relevance of Plato's words to
a far wider area, however, can be verified by any present-day visitor
to the region—if he can believe that the endless rocky, infertile vistas
once carried "plump soft parts" and rich forests.

As the Roman Empire entered an economic crisis period in the
third and fourth centuries A.D., aggressive land-clearing throughout
most of the dominions ground to a temporary halt. By then the only
extensive forests in the countries ringing the Mediterranean survived
in the mountains. Livestock, farmers, and the ship-builders of maritime powers like the Byzantine Empire, Venice, Genoa, and Cata-
lonia subsequently made heavy claims on the remaining forests of
what are now Turkey, Greece, Yugoslavia, Italy, and Spain.

The influx of people and livestock, and military activities associated with the Arab invasions of North Africa of the seventh through eleventh centuries, quickly multiplied pressures on the limited forest there. A further displacement of local residents by French colonists in the nineteenth and twentieth centuries forced more farmers into the mountains, necessitating a new wave of forest clearance. Early industries and railroad development also took their toll. Rapid population growth in the last five decades has multiplied the need for firewood, grazing areas, and croplands. Forests that once covered a third of the total area of Morocco, Tunisia, and Algeria had been reduced to perhaps 11 percent of the area by the mid-twentieth century. Though major reforestation programs are being attempted throughout North Africa, the losses still outpace the gains earned by new plantings.[10]

Central and Western Europe were heavily wooded in prehistoric times. With the advent of agriculture, land clearing by ax and fire increased, and deforestation proceeded slowly until, by the twelfth century, the substantial reduction in forested area was plain. Through the Middle Ages, European forests expanded or contracted with the ebb and flow of human populations affected by plagues, wars, and social reorganizations. Over most of central and western Europe, the spread of agriculture apparently peaked by 1300; the era of expanding arable land was followed by two centuries of stagnation and contraction.[11] In temperate central Europe, unlike the drier Mediterranean zone, forests were able to recover naturally when given the opportunity to do so.

Europe's lengthy economic recession drew to a close near the end of the fifteenth century, and with it ended the respite for forestlands. The acceleration of science and industry associated with the spread of the Renaissance from Italy brought not only a cultural flowering, but also a deflowering of many of Europe's remaining untouched forests. Expanding glass, mining, smelting, tanning, soap, and other industries required huge quantities of timber. Glass-makers obtained the potassium they needed from wood ash and their heat from charcoal or timber. The soaring production of iron and steel made particularly huge demands on woodlands for smelting. The manufacture of steel (an alloy of iron and carbon) long depended upon charcoal as its carbon source. The eventual adoption of coal rather than wood as a heat and carbon source for blast furnaces in the eighteenth and nineteenth centuries may be all that prevented the Industrial Revolution from totally swallowing Europe's woodlands. The colonial expansion of Europe further accelerated timber needs for ship-building, and the declining availability of suitable timber became an acute problem throughout the continent until 1862, when an American Civil War battle announced the age of iron-clad ships. For sound reasons, arable lands in flatter zones had been cleared for agriculture, but the forest cutters and charcoal producers of these early European industries habitually clear-cut one mountain slope after another without any consideration for regeneration.[12]

Although, as we shall see, the pernicious consequences of such indiscriminate tree destruction were then becoming apparent in Europe, and were sparking second thoughts about forest strategies, European conquerors competing for empires in the New World took environmentally destructive habits with them. Oversized herds of cattle, sheep, and goats brought from the Old World have helped destroy needed woodlands, create deserts, and degrade rangelands through much of South America, Mexico, and the southwestern United States ever since their introduction in the sixteenth century. Construction in the new Spanish cities of the Andean region and Mexico was a major cause of deforestation, much of it on erosion-prone slopes. A famous stand of cedars in the hills near Mexico City followed the fate of its ancient Lebanese relative: it was obliterated during the reconstruction of Mexico City after the Spanish conquest.

Even more damaging than city construction were the newly opened mines of Mexico and the Andes. Land denudation and subsequent erosion wrought by the quest for mineshaft timber and smelting charcoal soon became notorious in both regions. L. B. Simpson has described what happened in Mexico:

> Wherever a mining community was established a diseased spot began to appear, and it spread and spread until each mining town became the center of something like a desert. As early as 1543 the Indians of Taxco complained to Viceroy Mendoza that all the forests nearby had been cut down and that they were obliged to make a day's journey to get timber for the mines. The mountains of Zacatecas were once covered with a heavy forest. After four centuries they are now rocky grasslands where the ubiquitous goat has to scratch for a living. The same story was repeated at all the other famous mining centers of Mexico: Guanajuato, Ixmiquilpan, Zimapan, Pachuca, Chaucingo, Zacualpan, Temascaltepec, Tlalpujahua, Parral, and the rest.[13]

Psychological as well as economic needs apparently influenced the Spaniards as they cleared the hills of central Mexico. Accounts from the period suggest that the homesick colonizers cut trees partly so that the terrain would resemble that of their native land, Castile, which had been so denuded for centuries that, according to an old saying, a bird could fly over it and never find a branch on which to rest. Whatever the causes, the deforestation of Mexico was rapid. Early explorers estimated that from 40 to 50 percent of Mexico supported marketable timber; by 1950 that total had been reduced to 10 percent, and it has continued to fall.[14]

To the north, other European colonists moving inland from the Atlantic found one of the richest forests in the world. The settlers of the United States, generally mesmerized by the boundless expanse of American forests, and often charged with a frontier spirit oblivious to the delicacy of environmental balances, proceeded to raze large areas in their drive to forge a new civilization. Considerable clearance for farms and settlements was, of course, necessary and desirable, but many environmentally strategic forests were also destroyed. By the early twentieth century, a third of the country's forests had disappeared.

At this point conservationists and politicians began to worry as erosion, siltation, and flood problems—all linked to forest trends—began to mount. Gifford Pinchot, the doyen of American foresters, reportedly showed President Theodore Roosevelt a fifteenth-century painting of a beautiful, populous, well-watered valley at the foot of forested mountains in North China—and then a photograph of the same valley taken about 1900. "The photograph," as Walter Lowdermilk recounts, "showed the mountains treeless, glaring and sterile; the stream bed empty and dry; boulders and rocks from the mountains covering the fertile valley lands. The depopulated city had fallen in ruins. The President illustrated his message to Congress with these pictures and caused the establishment of the U.S. Forest Service for the protection of forest lands."[15]

Extensive, federally controlled national forests were subsequently created. The principles of sustained yield forestry, under which a long-term balance of timber cutting and regrowth is sought, and multiple use, which recognizes that forests serve a variety of economic, ecological, and recreational purposes in addition to supplying wood, were incorporated into Forest Service policy. While the effectiveness with which these ideals have been implemented has often been a source of controversy, the United States is protected against the fairly mindless, large-scale forest destruction of its first century of existence. In many regions, particularly the Southeast, once-devastated watersheds have been replanted.

Early American foresters had a rich body of knowledge and experience from which to draw. In the eighteenth and nineteenth centuries, forestry had emerged as a science in Europe, particularly in England, Germany, and France. Like many other intellectual breakthroughs, scientific forestry was born out of necessity: the need for ship timber and wood for industry had forced attention to forest-regeneration techniques.

The rampant clearing of mountain slopes in the Alps and Pyrennees produced a severe ecological backlash. Strange occurrences were noticed in northern Italy as early as the mid-fifteenth century, providing an object lesson in forest ecology. Smelters, which had long obtained their fuel from the adjacent slopes, began finding their summer operations interrupted as streams temporarily dried up, and the incidence of floods and deadly landslides also rose. A French prefect's report in 1793 summarized the problem well: "Around Grenoble, mountainsides have been denuded of their tree cover to such an extent that there are only barren rocks left. Each rain causes terrible damages. Rivers have no steady water flow any more. They are either low, or after rains, carry torrential amounts of debris, which devastates the lower fields and makes navigation near their junctions with the Rhone impossible. There are fewer wells and headwaters and less irrigation for the fields."[16] By the end of the eighteenth century, landslides and raging floods, known as alpine torrents, were a chronic threat in Austria, Germany, Italy, France, and Spain.

Sudden torrents overwhelmed farms, homes, and lives, but also sparked pioneering studies of the relationship between forests and environmental stability. In 1797 a French engineer named Fabre published his conclusion that the violent torrents were the result of deforestation in the high Alps. Contemporary and subsequent studies throughout Europe further established the role of people in undermining their own livelihood in the alpine valleys.[17] These findings eventually prompted massive reforestation campaigns throughout the Alps and Pyrenees in the nineteenth and twentieth centuries, which have ameliorated the problems of torrents and landslides in many areas, though the problems remain serious. Throughout most of Europe, in fact, the trend of deforestation has been halted and the forested area, most of it expertly managed, has gradually increased in the current century. Today, more than one-fourth of central and western Europe is forested. H. C. Darby, the chief historian of Europe's forests, observes that "Could the Roman legionaries tramp the countryside once more, they would be reminded only rarely of the dense shades they had encountered when they left the Mediterranean lands to build an empire."[18] Still, neither would they find the accelerating devastation of the landscape that characterized their homelands and much of Europe in their own time.

By the mid-twentieth century, human beings had reduced the world's original forested area by at least one-third, and perhaps by one-half. Using a loose definition that includes everything from desert scrub and lightly wooded Arctic tundra to luxurious jungles, 29 percent of the land surface of earth could be considered as forested in 1963 when the last global survey was undertaken.[19]

It is impossible to estimate how rapidly this area is now shrinking; most poor countries are fortunate if one good inventory of their forests has been carried out during the last several decades, let alone two or more to permit comparisons. Published national forestry statistics are often misleading. They usually cover only those wooded areas officially designated as "forest lands" by governments—and even the figures for these areas are sometimes grossly doctored. The United Nations researchers carrying out the 1963 *World Forest Inventory* discovered that as much as half the area reported as "forest land" in many countries was also labeled "unstocked"—generally a euphemism for partially or wholly denuded lands on which reforestation remains a hypothetical prospect.

In the industrial countries, ecologically vital forests are often reasonably well-protected, and in many areas agricultural progress since World War II has released considerable marginal farmland for reforestation. Where clearing continues, it seldom poses a major threat to farmlands and water cycles. Still, serious environmental disputes over forests continue to occur in these countries. A general concept like "multiple use," for example, can mask bitter clashes among competing interests. Priority uses of a forest for heavy grazing, as a productive timber source, as a public recreation area, or as a primordial preserve require quite different management strategies, and the choice of timber-harvesting techniques has both biological and esthetic ramifications.

In most poor countries, by contrast, the central issues posed by forests are more basic. In some countries the first step—the designation of adequate forest reserves—remains to be taken. Elsewhere the protection of such forests from farmers, wood poachers, grazing livestock, or officially licensed but unscrupulous wood merchants remains vital. And outside legally protected forests in many countries, the trees of the countryside are disappearing at a frightening pace with huge, if unmeasured, economic and ecological costs.

Areas that were densely settled long ago, such as the Middle East, parts of North Africa, the Andean region of South America, northern Ethiopia, and much of China, generally lost their stock of trees in ages past. But many other poor countries are, in this century, passing through the same stage of accelerated forest destruction that Europe, and later the United States, passed through in the last several centuries. For many, unfortunately, there seems little prospect that Western Europe's rather benign experience—ecological recovery after nature made its needs clear to human beings—will be the model for present-day transformations. The pressure of rapidly multiplying populations on rapidly depleting woodlands is too great; the financial, technical, and political forces working to reverse it too meager. As barren landscapes are created, there will be no New World to colonize and few virgin tropical forests to serve as foreign lumber sources. Like the Middle Eastern civilizations of antiquity, some countries are in the process of desecrating their landscapes to the point of no return. Once enough topsoil washes or blows away, the capacity of the land to support either trees or human beings is permanently stunted.

The evaluation of current forest trends in various regions depends in part on the perspective of the evaluator. Most of the Indian subcontinent, China, West Asia, North Africa, Central America and northwestern South America is depleted and forest-poor by any account. In all these areas, wood and its products are scarce and expensive; in all, the absence of trees has taken a heavy toll on the environment and its productivity for people. With the notable exception of China, the net loss of trees continues in these regions.

But in Southeast Asia, Central Africa, and the Amazonian heart of South America the situation is more equivocal. In these humid tropical zones, the lumberman sees great, untapped resources of valuable timber—and also a terrible economic waste as prime species are thoughtlessly felled by farmers in the hinterlands. The ecologist, in turn, sees whole ecological zones under seige, with untold species being driven to extinction, and harmful consequences for environmental stability as the forested area is gradually cut down.

Most of Western Europe has achieved a reasonable balance between the ecological need for forests and other land uses, but it is a balance maintained in part by large imports of wood and wood products from northern Europe, the Soviet Union, and tropical Africa. Japan, too, is now protecting its steeper slopes from deforestation, but imports a huge volume of timber from Southeast Asia. Only North America, the Soviet Union, and northern Europe have both great economic forest wealth and generally adequate ecological forest protection.

The two principal causes of deforestation today are land clearing for agriculture and wood gathering for fuel. A third cause is lumber harvesting for direct or industrial use, but, as a source of deforestation, this is far less significant on a global basis than the other two. Much of the world's timber and forest-products industry manages its available forests on a sustained yield basis. In some of the areas where timber concessionaires, local or foreign, are often less considerate of the future, such as in Southeast Asia and tropical Africa, the principal negative consequence is frequently not land denudation—vegetation of some sort springs up quickly in the humid tropics—but rather the economic loss as the more valuable and usable species disappear. In Nigeria, for example, where forests have been overexploited for decades to meet local and foreign commercial wood demands, a leading forester fears that the country will "experience a degree of timber famine before the end of the century."[20] In areas of low rainfall, or on mountain slopes, irresponsible timber practices can cause severe ecological problems, as they are now in the mountains of Pakistan and Afghanistan. A layer of fertile topsoil that took centuries to build can wash away in one storm when land is completely exposed to the elements.

Since the Neolithic revolution, more forests have given way to farms and pastures than to any other use, and the spread of agriculture is still the biggest single cause of forest clearing today. Most of the farmland in virtually every nation was once at least modestly tree-covered, the main exceptions being irrigated desert lands and dry prairies. This shift from forests to farms, usually carried out by people who badly need the land's produce, is generally desirable; but if clearance occurs on ecologically strategic slopes, or on soils unsuited to agriculture, it becomes self-defeating. Even today, considerable potentially arable land remains forested in parts of Latin America, Africa, and Southeast Asia, though its extent is frequently exaggerated. The planned conversion of a good share of this area to cropland over the next few decades may be necessary to meet human needs, whatever the biological risks and costs in plant and animal species driven to extinction.

Unfortunately, however, the spread of cultivation is frequently more chaotic than rational. Colonists hungry for land as well as for food, sometimes moving under the sponsorship of governments unwilling to confront the need for a redistribution of proven farmlands, carve plots out of the forest where almost nothing is known of soil conditions. Much land colonization, say analysts of the U.N. Food and Agriculture Organization (FAO), "is indiscriminate . . . an ill-advised use of the land. It is merely a process of trial and error. Very often the chosen forest land cannot support permanent agriculture. When soil fertility is lost, cultivation is abandoned and the land is often grazed. The bare soil will frequently return to forest, unless, as is often the case, it is first destroyed by erosion."[21]

Settlement failures like these are frequent in the tropics, but perhaps even more harmful to life-support systems is the upward creep of cultivation onto forested mountainsides that should, for both economic and ecological reasons, remain tree-covered. Apart from the tropical rain forests, the last extensive forests left in many poor countries are on the steep slopes and sometimes inaccessible reaches of mountains. Nearly everywhere agricultural activity has traditionally been heaviest on the plains and valley floors, and with good reason, for severe erosion usually ensues when population pressures force farmers onto steep hillsides. The deteriorating environments of the Himalayas, the Andes, and the East African highlands (examined in Chapter 5) are dramatic examples, but the problem exists wherever sloping lands border crowded farm regions.

Clearing and farming these slopes is a precarious and usually futile business; crops are easily washed away, and erosion often necessitates the abandonment of a plot within two or three years. But any possible net gains for the individual are usually heavily overshadowed by the social costs of hillside deforestation. As the alpine torrents revealed to Europe centuries ago, people strip mountainsides of trees at great risk to their own well-being. To take one example, the Indonesians are now learning the same lesson in Java: as the clearing of hillsides by farmers accelerates, so does the load of silt that clogs Java's indispensable irrigation systems and destroys the usefulness of its reservoirs.[22] Dozens of similar examples tell the same story.

As desperate farmers, aided by wood merchants and firewood gatherers, clear more and more hills in the Himalayas, the incidence and severity of floods in the lowlands below is rising. The lives of twenty million people in India are directly disrupted by flooding in the average year, and when the rains are particularly heavy, as many as fifty million people are so affected. Monsoon rains rush off tree-shorn slopes, and the increased load of silt carried by the Indian subcontinent's rivers raises their beds so that annual floodwaters are spread outward more rapidly. Writing in the late sixties, a World Bank study group noted that high floods on the Indus plains of Pakistan have occurred more frequently during the last twenty-five years than over the previous sixty-five years.[23] In August, 1973, a flood considered by many to be the worst in Pakistan's history flooded nearly two million hectares of standing crops and ten thousand villages.

Flood problems are similarly accentuated by hillside deforestation in areas of eastern India, Thailand, the Philippines, Indonesia, Malaysia, Nigeria, Tanzania, and many other countries.[24] Though hundreds of millions of dollars are spent around the world each year on engineering projects like dams and embankments to control floods, engineering "solutions" can address only the effects, not the causes, of basic ecological imbalances. Floods are an incontrovertible and natural fact of life, but as watersheds are denuded and populations are crowded into flood-prone areas, the annual worldwide destruction of crops, homes, and equipment by floods is escalating.

In the tropical zones of Africa, Asia, and Latin America the dominant form of agriculture is shifting cultivation, whereby plots are cleared, farmed for a few years until their fertility wanes, then abandoned. This practice annually accounts for far more felled trees than the spread of permanent agriculture. In Asia, for example, the FAO estimates that 8.5 million hectares of forest are cleared annually by shifting cultivators; a woodland the size of Portugal is cleared *each year*. An estimated five to ten million hectares are cleared annually for agriculture in Latin America, mostly for temporary farming.[25]

When the ratio of humans to forest is low enough to permit the lengthy abandonment of these lands, woody vegetation soon re-establishes itself. Regeneration begins especially quickly in rain forest areas such as the huge basins of the Amazon and Congo Rivers. As the number of cultivators grows, however, so does the cleared area, until finally whole forests are threatened, as is the fertility of the soils. Successive aerial surveys of the dense rain forest of the Ivory Coast showed a reduction in forested area of 30 percent between 1956 and 1966.[26] Migrants from land-short regions in the north and neighboring African countries follow new logging roads into the forest to join local residents in clearing new farms. Too-frequent clearing in the humid tropics reduces precious high forests to thick shrubbery of little economic value and, in some cases, to tough grasslands difficult and costly to reclaim for farms or any other economic use. The loss of high forest to the tenacious grass *Imperata cylindrica* following temporary cultivation is a major problem in the Philippines, Indonesia, and parts of Africa.

The combined impact of shifting cultivators, ranchers, and logging industries has placed the world's tropical rain-forests in jeopardy. The problem is not usually that tropical rain-forests are being replaced by bare ground, though this is a looming threat in few areas, but that an incredibly diverse ecosystem, about which we still know very little, is being radically transformed at an amazing pace. Paul W. Richards, a leading world authority on tropical rain-forests, fears that the tropical-forest ecosystem as we have known it will virtually disappear from the face of the earth by the end of the twentieth century.

The costs of this decimation are impossible to measure in economic terms, but they are profound. As Richards recently wrote, "much of the plant and animal life of the tropics may thus become extinct before we have even begun to explore it . . . a vast realm of potential human experience may disappear before there is even a bare record of its existence."[27] The preservation of some limited tropical-forest areas in their original rich state is likely to be one of the major global conservation issues of the next decade.

At least half of all the wood cut in the world each year is burned as fuel, a practice whose many social and ecological ramifications are explored in Chapter 6. Unfortunately the accelerating depletion of world oil and gas reserves, and the high cost of alternative energy sources, guarantee that this particular demand for wood will not abate in the years ahead. In the huge semi-arid zones of Africa, Asia, and Latin America, which were only lightly wooded to begin with, fuel-wood gatherers are generally the chief agents of deforestation, and contribute to the creation of desert-like conditions. Throughout the remainder of these continents, the inescapable and expanding need for cooking fuel is responsible for the depletion of tree stocks both on and off the forest reserves. Where trees are plentiful, families may gather their own firewood, but where they are scarce, and around larger cities, a profitable trade in firewood or charcoal flourishes. Few of the fuel merchants maintain a stock of trees on a sustained-yield basis; instead, they obtain their basic raw material as a gift of nature. At present no one holds them accountable for the heavy costs to society of their business practices.

Growing shortages of trees for fuel, lumber, newsprint, and ecological protection have prompted many governments to recognize the need for reforestation. The annals of the past few decades are littered with ambitious national forestry plans unfulfilled by governments either unwilling or incapable of backing their schemes with sufficient money and political commitment. To take one notable example, India in 1952 adopted a National Forest Policy which explicitly recognized the importance of forests to society. It called for nearly a doubling of forest lands to eventually cover 33 percent of the country, a goal whose implementation would ameliorate much suffering in India. By 1975, India had designated 23 percent of its territory as forest land, but high officials privately admitted that little more than half of that area deserves the title. India's wood famine continues to pauperize both its people and its environment.

A more encouraging effort at forest rehabilitation is underway in China. Soon after assuming power, the new Chinese leadership recognized the futility of undertaking the needed agricultural development efforts without also giving more basic attention to the country's barren, erosion-prone landscape. Moreover, what was probably the world's most acute scarcity of forest products severely constrained overall national development, especially communications. The spread of the vital railroad system, for example, was hampered less by a shortage of steel than by the paucity of timber for ties.[28] Scarcities of fuel wood, paper, power-line poles, construction lumber, and timber braces for coal mines have all retarded China's economic development.

In the 1950s, ambitious though probably unrealistic plans were made for mass reforestation, to be implemented mainly by the rural communes. The leadership hoped to raise the forested area to 20 percent of the country by the late 1960s, and 25 percent by the 1980s. Not only were mountains and marginal or abandoned farmlands to be planted, but a "great green wall" of shelterbelts, reminiscent of China's famous Great Wall, was to encircle the huge Gobi Desert to halt the march of the "sand dragon."

Tens of millions of workers have since been mobilized to plant trees each year in the slack seasons between crop harvesting and sowing. The effort expended has been stupendous, though the results have sometimes been disappointing. By 1963, according to Chinese officials, about seventy million hectares—nearly 7 percent of the country—had been replanted, but in many areas the survival rate of the newly planted trees was, in this early period, below 10 percent.[29] Untrained labor, poor maintenance, mutilation or even uprooting of newly-planted trees for fuel, and frequently inhospitable growing conditions combined to corrode the efficiency of the mass planting efforts. By the mid-sixties, official emphasis had been shifted away from massive annual new plantings to careful maintenance within planted areas.

Accurate figures on the extent and success of China's reforestation programs are not available, but accessible written and visual evidence indicates that the progressive deterioration of more than five millenia has been reversed, and that the forested area is slowly increasing, though wood scarcity remains severe. As European authors who visited China in the mid-sixties put it, "The reforestation of China, however incomplete as yet, is now well-launched and definitely a fact. There is little doubt that this constitutes one of the major cultural accomplishments of our times, and the effect, both on the environment of China and on the development efforts of other countries, particularly wood-poor and ecologically disinherited nations, will not fail to manifest itself."[30]

3. Two Costly Lessons: The Dust Bowl and the Virgin Lands

A CLOUD OF DUST thousands of feet high, which came from drought-ridden states as far west as Montana, 1,500 miles away, filtered the rays of the sun for five hours yesterday," reported the *New York Times* on May 12, 1934, with the paper's characteristic sangfroid. The account continued: "New York was obscured in a half-light similar to the light cast by the sun in a partial eclipse. A count of the dust particles in the air showed that there were 2.7 times the usual number, and much of the excess seemed to have lodged itself in the eyes and throats of weeping and coughing New Yorkers."

Washington, D.C., the nation's capital, was overhung with a thick cloud of dust denser than any seen before. The whole eastern coast of the United States appeared to be covered with a heavy fog —a fog composed of 350 million tons of rich topsoil swept up into the transcontinental jet streams from the Great Plains, half a continent away. Even ships three hundred miles out in the Atlantic Ocean found themselves showered with Great Plains dust.

As the winds subsided the particles settled, covering half the nation with a thin layer of grit. In New York, Chicago, and Kansas City, businessmen expecting to get rich from their investments in Great Plains wheat or cattle empires saw their prospects dissipate first-hand, as their topsoil seeped around doors and windows to settle on top of their shiny office desks. The United States was learning an expensive lesson in ecology.

shiny office desks. The United States was learning an expensive lesson in ecology.

The Great Plains stretch northward from Texas through Montana and the Dakotas into Canada, including parts of ten states. Before the European invasions of the sixteenth through nineteenth centuries, these grass-covered, wind-swept plains were occupied by nomadic Indians and great herds of buffalo. Though Indian-set fires pushed back the forest in some areas and may have altered the grass species in others, land use before the nineteenth century was generally well-adapted to the region's environmental constraints.

The soils of the plains tend to be rich; their full exploitation is limited by water shortages. Throughout much of the region, grains can be grown in the wetter years, but when the rains fail—as they do periodically, usually in clusters of drought years—unirrigated crops can be devastated. And in no other region of the United States is the prevailing wind velocity so high. From both economic and ecological viewpoints, most of the Great Plains is better suited to grazing than to farming. Yet several times over the last century, prolonged periods of good rainfall have encouraged a counterfeit optimism about the capacities of the Plains among newcomers, and even among many long-time residents. Humans have repeatedly extended their farms and expanded their cattle herds beyond safe levels during the wet years, to be brought rudely back to reality when the drought cycle returned.

Perceptions of the environmental potentials and limitations of the Plains have varied with the climatic trends of each decade. They have also been actively influenced by shortsighted commercial interests and even by national moods. The semi-arid grasslands of the west-central United States were not always known as the Great Plains; early in the nineteenth century, the region was known as the Great American Desert. If this label exaggerated the area's aridity, it was probably less inaccurate and certainly less dangerous than the myth of fecundity that followed it. The term "Great Plains" came into general use only after the American Civil War of the 1860s, in part because of the deliberate propaganda of railroad companies and land speculators who wished to sell off their holdings and also to see the Plains settled, creating business for the trains. Newspaper adver-

tisements designed to lure Easterners westward pictured lush corn
fields, rich pastures and prosperous families, and helped to fabricate
an image of the Plains as a garden spot. The claims of profiteers
coincided conveniently with the prevailing national mythology of the
American West as a potential agricultural empire, a place where the
good life would come easily to those enterprising enough to move
there and take advantage of it.[1]

And settle the Plains they did. In the 1880s, the first wave of
farmers moved into the region, their barbed wire fences marking the
closing of the earlier era of cattle empires based on unfettered access
to a vast open pasture. The federal government's offer of 160-acre
homesteads sounded attractive indeed; homesteaders in moister
woodlands and grasslands to the east had prospered with the same
acreage allotment. It was not long before the first drought revealed
the impossibility of supporting a family on a parcel of that size in the
Great Plains. Farmers continually resorted to planting grains on fields
that needed a year-long fallow period to permit the accumulation of
moisture in the soil, and there was inadequate space left for livestock.
Poor rainfall in the 1890s drove many of the homesteaders and other
settlers away, forcing them to sell their land to neighbors or speculators. Others remained to become tenant farmers on the holdings of
absentee landlords. After another drought in 1910 brought ominous
clouds of dust and drove more farmers off the land, two new forces
began to influence developments on the Plains. The tractor, the
combine, and other power machinery permitted a farmer to plant and
harvest a larger area than ever before. For many, the costly new
machines were a mixed blessing; farmers went heavily into debt to
buy them and then were compelled to maximize crop production in
order to obtain enough cash to meet their payments. Second, as
World War I disrupted European agriculture, international wheat
prices soared, enticing farmers to convert more and more pastures
into wheat fields.

In the 1920s, which environmentalist Paul Sears has labeled the
"cloud nine decade" of the Plains, the extension of wheat farming
continued. Smaller farmers desperately trying to pay off their debts
planted as much wheat as they could, even after the price plummeted
in the postwar period. Big-time mechanized wheat operations, as well
as huge cattle ranches, attracted investments from Wall Street, and

in some cases from as far away as Europe. "Suitcase farmers" plowed and planted huge holdings with their tractors in the fall, retired to Florida or California for the winter, then returned for the quick harvest and sale of their crop in the summer; if prices were too low, they did not bother to return, instead leaving their dying crops to the winds.[2]

Though periodic droughts and strong winds had always plagued the Great Plains, in the early thirties their effects on the land became visibly worse. The Plains normally turned brown when the rains failed, but the intertwined roots of the hardy native grasses shielded the earth from the wind. Now, however, on huge areas of former grassland plowed under for grain planting, the gales of the Great Plains lashed against the unprotected ground. Soils desiccated by drought ceased to support plant life; instead, their lighter particles became dust in the air, while the heavier particles rolled about until they formed snow-like drifts of sand that smothered fields and machinery. By the summer of 1933, after two successive years of drought, millions of acres planted to grains were not harvested because the plants were too shriveled to repay the effort. Cattle were dying on the parched rangelands, and thousands of farm families found themselves broke and able to fight starvation only by signing on for government make-work programs. Others gave up, or found themselves displaced by the tractors landlords were purchasing in growing numbers. Many of these refugees made the westward trek, so well-chronicled by John Steinbeck in *The Grapes of Wrath*, to another mythical land of bounty—California.

In the spring of 1934, as hot, dry winds introduced another failure of the grain crops, the situation reached an awesome climax. On April 14, a month before the storm that inundated the eastern half of the country with dust, families of the Plains learned in no uncertain terms that their region was entering a new era. Winds from the north swept through Kansas and eastern Colorado, skimming soil off the roots of crops as they went and building into a massive ebony cloud that avalanched southward. The cloud was sufficiently thick to cause hours of eerie total darkness at midday, and the litter of dead birds and rabbits in its wake revealed the dangers of breathing the palpable air. People in its path could only tie handkerchiefs around their faces to help ease the pain of breathing.

As the dust storms continued intermittently over the next several years, cases of pneumonia, nicknamed "dust pneumonia," were to double in some counties, and other respiratory ailments increased as well. Many found the dark color of the dust storms of the thirties to be especially frightening, and with good reason. Soil scientists determined that it was a disproportionately high share of the topsoil's richly fertile organic matter that gave the clouds their black hue.[3]

President Franklin D. Roosevelt chose drought as the subject of his first "fireside chat" with the nation in the 1936 presidential campaign. His account of the devastation he saw in a tour of the Plains could easily have been drawn from 1973 news accounts of the West African drought: "I talked with families who had lost their wheat crop, lost their corn crop, lost their livestock, lost the water in their well, lost their garden and come through to the end of the summer without one dollar of cash resources, facing a winter without feed or food—facing a planting season without seed to put in the ground."[4]

The rain of dust over half the nation, the vivid tales of human suffering, and the images of a potential new desert being created by Americans in the heart of their own country captured national attention far more effectively than the urgent warnings that had been offered for decades by scientists. Paul Sears has recounted what surely was one of the most convincing political lobbying efforts ever undertaken in Washington. The land argued its own case as dust from a thousand miles away visibly filtered into the congressional hearing room where Hugh H. Bennett of the U.S. Department of Agriculture, then one of the world's leading soil conservationists, pleaded for new funds and programs to protect the nation's topsoil.[5]

The Dust Bowl of the thirties catalyzed a major turn in the ecological history of the United States. A century and a half dominated by the empty legend of endless frontiers and boundless resources had left much of the country's land in deplorable shape. Indeed, many experts agreed with Walter Lowdermilk, a noted colleague of Bennett's, when he observed in 1935 that "America has been developing desiccated and unproductive lands more rapidly than probably ever before occurred." Research in the 1920s had shown at least eighty million hectares in the country to be suffering from accelerated erosion, with twenty million formerly productive hectares already abandoned.[6] Though erosion was by then a critical national problem, it took the spectacular collapse of an entire region to spur the political leadership into action on a meaningful scale.

The unprecedented dust storms were soon followed by the formation, in 1935, of a Soil Conservation Service, with Bennett as its first chief. Federal promotion of soil conservation practices in the Great Plains, and soon in every state, helped to arrest and, in many areas, reverse the accelerating trend of soil degradation which many at the time believed was jeopardizing the nation's survival. In the Great Plains, a federally appointed Great Plains Committee studied the state of the land and recommended solutions. The committee's eloquent description of trends in the Plains, issued in 1936, is a classic statement of the continuing plight of enormous semi-arid regions around the world today:

> Current methods of cultivation were so injuring the land that large areas were decreasingly productive even in good years, while in bad years they tended more and more to lapse into desert. . . .
>
> The steady progress which we have come to look for in American communities was beginning to reverse itself. Instead of becoming more productive, the Great Plains were becoming less so. Instead of giving their population a better standard of living, they were tending to give them a poorer one. The people were energetic and courageous, and they loved their land. Yet they were increasingly less secure on it.[7]

As rains returned to the Plains in the late 1930s, soil conservation officials urged upon farmers a variety of practices to restore and preserve the land's utility. Millions of hectares of especially vulnerable cropland were returned to pasture. Fallowing, supplemented by the calculated use of crop stubble and plant residues left on the unplanted fields to protect them from the wind, was advocated to conserve moisture and thus ensure a better crop in the years of planting. Strip cropping—the alternation of bands of grain with bands of other more soil-protective crops or grasses—helped cut wind erosion, as did the wholesale planting of trees in windbreaks between fields. Terracing of fields, and plowing along the contour of the land, were encouraged to reduce erosion by water. Herd sizes were limited to prevent the destruction of pastures by overgrazing.

Drought plagued the Plains again in the early 1950s, visiting some

areas with even greater severity than two decades before. Improved
farming practices helped prevent the large-scale regional debacle of
the 1930s, but land damage was widespread enough to provoke another federal inquiry. This time a special new program for the Great
Plains was established, under which the federal government provides
financial assistance to farmers who wish to implement conservation
measures.

Surging world grain prices in the mid-1970s, and consequent
all-out production efforts by U.S. farmers in 1974 and 1975, have
inculcated new fears for the land and provoked strong warnings from
agricultural scientists. Government surveys show that of the nearly
four million hectares of former pastures, woodlands, and idle fields
converted to crops nationwide in late 1973 and early 1974, over two
million hectares had inadequate conservation treatment. The average
loss of topsoil to water and wind on the latter lands in 1974 was
twenty-seven tons per hectare, more than double the twelve tons
considered "tolerable" by government soil conservation officials (a
figure still above the rate of natural soil regeneration in many regions). In the southern portion of the Great Plains, the surveys
showed, the year's soil loss on twenty thousand hectares of newly
planted land ranged from thirty-four to 314 tons per hectare.[8]

In a report to the U.S. Senate in early 1975, a committee of
leading experts on the private Council for Agricultural Science and
Technology warned that "the farm community may be creating
another dust bowl." Despite the substantial progress of the last four
decades, they noted, several recent trends could lead to another surge
of land devastation should prolonged drought reoccur on the plains.
In response to high wheat prices, strip cropping and fallowing have
often been neglected in favor of continuous wheat cultivation. Together with low cattle prices, high grain prices have also encouraged
a limited shift of rangeland into crops in drier, riskier areas. Many
farmers, faced with a growing weed problem on continuously planted
wheat fields, have returned to once-abandoned plowing techniques
that bury the weeds—but that also leave the soil surface bare of
wind-resistant plant residues. The leveling of sandy hills to permit
new center-pivot irrigation techniques, the committee observed,
often exposes easily blown barren sands. Finally, high land and com-
modity prices are promoting a reduction in the area planted to trees
as windbreaks.[9]

Whether the market conditions encouraging these dangerous trends will persist, and whether the alarms being sounded will spur corrective actions, remain to be seen. Meanwhile, some basic lessons emerge from the history of the Great Plains. Hearteningly, the human ability to reclaim devastated semi-arid regions and, through proper management, to exploit them in a sustainable fashion has been proven. Less encouraging is the continuing widespread erosion in the Plains and the lurking possibility of another Dust Bowl emerging despite the ordeals of the past and the ready availability of conservation technologies. Apparently, neither the hard-bought and soon-forgotten lesson of the Dust Bowl, nor the unregulated forces of the free market, is sufficient to safeguard the soil.

Ecological lessons seem to travel the oceans slowly. The Soviet Union, like the United States, has learned through experience that efforts to stretch natural systems too far will eventually backfire. Nikita Khrushchev inherited some difficult agricultural choices when he assumed leadership of the Soviet Union in 1953. Food production in the previous year had barely been above the pre-revolutionary level of four decades earlier, while growing numbers of consumers were clamoring for a better diet. For decades, the country's agricultural sector had been drained of capital to finance the building of factories, but by now it was apparent that grain production had to be boosted rapidly and massively.

Raising the yields substantially on existing farmlands, Soviet leaders calculated, would be tremendously expensive. Multi-billion-dollar irrigation schemes and costly fertilizer factories would have to be built, channeling resources away from the drive to catch up with their Cold War competitor nations in heavy industry and the production of consumer goods. Instead, they looked to the east. In northern Kazakhstan, western Siberia, and eastern Russia, they saw vast grasslands untouched by the plow, and millions of hectares of established farmlands intentionally left idle each year by farmers. Perhaps plowing these "Virgin Lands" and reducing these wasteful fallow periods would make it possible to multiply the country's grain production at comparatively little cost. . . . The Soviet weather roulette, with its frequent droughts in one region or another, meant there was some inherent risk, but even two good harvests out of three would repay the investment many times over.

A gigantic program to farm the Virgin Lands and ameliorate the Soviet Union's chronic food problem quickly and cheaply was vigorously undertaken. Between 1954 and 1960, hundreds of thousands of enthusiastic settlers brought forty million hectares of new land, an area over three times the size of England, under cultivation. They faced great hardships, often living in tents and using inadequate equipment, but many took to heart Premier Khrushchev's exhortation to work hard, and the initial results were impressive. National grain output climbed by 50 percent over the six-year period, mainly because of these additions to the country's cultivated area.[10]

By 1958, however, some farmers in the fields were openly questioning the government's strategy. The directors of one state farm sent their protest to *Izvestia*, the official government newspaper: "Why are we given sowing plans that require us to cut the ground from under our feet—to use a figure of speech—to fulfill them? Can it be that the officials of the republic organization are not aware that if grain crops are planted in the same soil for the fourth year running, that soil will be exhausted?" Although the planted area had increased fourfold on their farms, the directors complained, they had been instructed to hold fallow 429,000 hectares fewer than they had before the expansion began. "Perennial grasses," they warned, "have been eliminated completely."[11]

Alarmed by mounting wind erosion in the Virgin Lands, A. Barayev, director of the Grain Farming Research Institute at Shortandy, experimented with farming methods quite different from those recommended by Moscow. Keeping a third of the farmland out of crop production each year, he figured, would bring the greatest total grain production over the long run. Moisture could build up in soil as it rested in "clean fallow," encouraging higher yields in the times it was cropped, and reducing the chances of severe wind erosion of desiccated soils. In contrast, the official "crop fallow" system specified that corn, rather than grass or leftover stubble, was to be planted in what were supposed to be rest years for the land between wheat plantings.

Khrushchev advocated deep plowing that would remove the stubble left from previous crops, thus allowing traditional planting equipment to proceed more quickly, and planting to be completed earlier in the season. But Barayev preferred shallow plowing with equipment designed to leave stubble on the ground. Stubble, which he called the soil's "armor," would prevent the vicious winter gales from blowing the snow cover off the ground. Otherwise, he observed, the soil tended to freeze to great depths, and, when snow melted in the spring, it would run off the surface rather than soaking in for later use by crops. In spring and summer, stubble would help protect the soil from the persistent winds, and it was an essential guard of topsoil on fallowed fields. The early planting Khrushchev urged would reduce the chance of widespread crop losses to an early snow at harvest time, such as occurred in the fall of 1959, but this practice ensured major crop and soil damages whenever the unreliable May rains came late—a far greater threat to long-term productivity, in Barayev's eyes. Finally, Barayev saw the alternation of belts of perennial grasses with belts of annual grains as essential to holding down wind erosion, though this strip cropping meant sacrificing grain output in any given year.[12]

With his unorthodox ideas, Barayev was not popular among the national leadership in the early sixties. Those in power found more appealing the methods of A. Nalivaiko, director of the Altai Institute, who supported early planting and crop fallowing rather than clean fallowing. Premier Khrushchev's preferences were clear when he spoke to Virgin Lands agricultural workers in late 1961:

A conversion to the correct system of farming is of great importance both for the Virgin-Land areas and for the country as a whole. At the 22nd Party Congress I spoke of the advisability of having clean fallow wherever it is needed. There is a dispute on this matter between the Altai Research Institute, of which Comrade Nalivaiko is director, and the Grain Farming Research Institute, directed by Comrade Barayev. I must side with Comrade Nalivaiko in this dispute.

You yourselves understand that this is not a matter of personal sympathy. The system recommended by the Altai Institute is more effective and gives better results. I have probably heard Comrade Nalivaiko speak four times. I have listened to his arguments, and I think he is on the right track.

. . . What is our approach to clean fallow? If we are speaking from the point of view of farm development, the answer would be that the less clean fallow there is, the better. If there is no clean fallow and a good harvest is obtained, this is best of all. I call upon you to study and decide for yourselves to what extent this is possible. . . . But, comrades, take risks, because it is not always possible to achieve the desired goal without taking risks.

. . . If you can prove in practice that you can obtain more grain per
hectare of plowland than Comrade Nalivaiko gets, if you win the competition, Comrade Barayev, we will quickly turn to you. . . . I do not think you
will win.

Put Comrade Barayev into conditions of capitalist competition and his
farm, with its present system of plantings, probably would not survive. Could
he ever compete with a large capitalist farm if he keeps 32 percent of his
plowed land in clean fallow?[13]

The new state farms of the Virgin Lands followed the premier's
advice and took risks, but events of 1963 revealed the heavy odds
against this gamble with nature. As a dry spring became a dust-filled
summer, desolation crept over the Virgin Lands. The Virgin Lands
had no John Steinbeck to record indelibly the social consequences of
the emerging dust-bowl conditions, but the direct agricultural consequences were more easily registered. Crops on three million hectares
were lost altogether to the drought.[14] Yields were slashed throughout
the region, and the normally savage winds carried precious topsoil,
now dehydrated and easily torn from the earth, off the farms.

The dramatic crop failures and land devastation of 1963 were a
serious challenge to the national leadership. Questions about the
wisdom of Khrushchev's agricultural strategy grew louder and, with
supporting evidence readily visible in the fields, more pointed. In a
May, 1963, issue of *Izvestia*, Barayev outlined the steps needed to
combat the erosion menace—practices that the premier had derided
two years earlier.

Other analysts were adding up the costs of unsound farming
techniques, costs that had begun accumulating well before the 1963
debacle. In one district of Kazakhstan, for example, up to 180,000
hectares of grain had annually been lost to the wind between 1955
and 1959. In 1960, 450,000 hectares were decimated; in 1961,
700,000; and in 1962, 1.5 million. The total rose still further in the
next year. Not surprisingly, as the losses of planted crops climbed, the
yields on fields still worth harvesting fell to dismal lows. The government had originally projected Kazakhstan's average grain yield to
reach two tons per hectare by 1963, but the yields peaked in 1956
at only 1.1 tons, and plummeted to 0.64 tons in the exceptional
drought year of 1963. F. A. Morgun, a senior Soviet agricultural
official, later calculated that in the 1962–65 period a total of seven-

teen million hectares were damaged by wind erosion in the Virgin Lands, with four million of these lost to production altogether.[15]

In 1963, Premier Khrushchev appointed a commission to study the year's crop failure, but its findings were never made public. In December of that year, Barayev, while retaining his research post, was elected to the high position of deputy of the Supreme Soviet. Three months later, he was invited to address the national Communist Party's Central Committee—a tangible indication that support for his ideas was accumulating among powerful national leaders. Khrushchev himself admitted, in an early 1964 interview with an Italian publisher, that the exhaustion and erosion of the Virgin Lands did indeed require the return of some portions to grazing land or fallow. But later in the year he was again exhorting the state farms to engage in all-out production.[16] In October, 1964, Khrushchev was deposed by his fellow party leaders, his disastrous agricultural policies part of the mosaic of grievances that led to his political demise.

Khrushchev's successors proved more erosion-conscious. Beginning in 1965, new machinery designed to leave crop stubble on the ground was employed and the annual fallow area was increased. Tree planting was stepped up to create windbreaks on the bleak plains of the Virgin Lands.

The dreams that originally inspired the Virgin Lands campaign will never be fulfilled, and erosion will always be a serious menace in this region's inhospitable conditions. Once human practices were adapted to the environment, however, the program did evolve from an embarrassing liability to a moderate success. These new farmlands provide the Soviet Union with an important back-up harvest that can ameliorate the impact of the periodic crop failures on the more productive western farmlands, though massive agricultural investments elsewhere in the country have also proved necessary. And, as the world learned in 1972, and again in 1975, massive grain imports from abroad are also necessary in years when nature is uncharitable to the Soviet Union.

4. Encroaching Deserts

Hot, dry regions cover more than one-third of the earth's land surface. In atlases and daydreams they usually appear as vast blank areas, but in fact they encompass a variety of ecological zones, ranging from lifeless seas of sand dunes to lush irrigated oases. Where human ingenuity has devised means for diverting river waters onto them, desert lands have frequently become extraordinarily productive and densely populated. Indeed, desert civilizations that harnessed the flow of such rivers as the Tigris, the Euphrates, the Indus, and the Nile dominated early human history.

In less richly endowed regions of the central Sahara or Saudi Arabia, one can travel through hundreds of miles of sand without seeing a sign of life. There are spots in the Atacama Desert of western Chile where no rainfall has ever been recorded. Most lands that might be considered arid or semi-arid, however, are between these two extremes. They include the thinly vegetated desert fringes that sustain only nomadic populations and their grazing animals, and the larger areas comprising much of central Africa, Asia, and Latin America that support a settled population, crops, and livestock. The single defining factor linking all these zones is low, often unreliable rainfall that limits the level and kinds of animal life and vegetation they can support.

Less than half the world's drier zones are unproductive deserts for climatic reasons, but usable land is being converted to waste by the influence of man and livestock. When water supplies are limited, a population density that seems low on paper can be overwhelming on land. Deserts are creeping outward in Africa, Asia, and Latin America, and, worse, the productive capacity of vast adjacent semi-arid regions is being shrunk. About sixty million people in the poor countries live directly on the interfaces of deserts and arable lands, and the basis of their livelihood is in jeopardy.[1] Once lost to desert sands, land is reclaimed for human benefit only at enormous cost.

Deserts have held sway over outsiders frequently over the millenia. The splendors of ancient Egypt lured Roman armies and emperors. For both Christians and Moslems desert areas are the Holy Land. In the early 1970s world attention has again been riveted on arid lands—on some because of new-found wealth, and on others because of human and ecological catastrophe. As word spread of a great drought and famine south of the Sahara Desert in Africa, little-known countries such as Chad and Upper Volta suddenly appeared in front-page newspaper stories. The hollow faces and stick-like limbs of starving humans haunted television screens around the world. Belatedly, a huge international relief effort got underway, eventually eliminating most overt starvation in the afflicted areas. Even so, hundreds of thousands of refugees, mostly proud nomads never before so humbled, poured into West Africa's larger cities begging for sustenance. And still there were deaths—perhaps hundreds of thousands of them.

A term unfamiliar to most, the Sahel, soon became common journalistic coin. Actually a geographic description of the narrow band bordering the Sahara, derived from the Arabic word for "border," in popular usage the term Sahel has often been applied in a general way to the six West African countries that faced drought from 1968 through 1973: Mauritania, Senegal, Mali, Upper Volta, Niger, and Chad. In fact the drought affected not just the Sahelian zone, but also the broader Sudanian zone to its south, which includes most of these six countries, and parts of many others in central Africa as well. Then, even as the West African tragedy was unfolding, similar droughts and famines struck the countries to the east, including Sudan, Somalia, Ethiopia, Kenya, and Tanzania—droughts that persisted in some areas in mid-1975.

As awareness of the misery and dislocation of millions of Africans grew, so did a realization that something deeper was occurring than just a hiatus in the area's always low rainfall. Those studying the history and ecology of the Saharan fringes quickly discovered that the ecological calamity triggered by the drought had been stewing for decades. Its roots lay in social and economic patterns incompatible with the region's environmental limitations; the fundamental predicament remained when the rains finally returned to most of the Sahel in 1974.

Disasters in the desert are nothing new; droughts and crop failures have always plagued arid lands, as Joseph recognized when he advised the Pharaoh to set aside grain reserves in ancient Egypt. But both the scale of suffering when the rains fail, and the scale of destructive human pressures on delicate arid zone ecosystems, are reaching unprecedented proportions in the Sahel and many other desert regions. Populations are, in effect, outgrowing the biological systems that sustain their lifestyles, and it is an open question whether their ways will change in time, or their life-support systems will disintegrate irretrievably.

Many deserts are spreading, but the terms used to describe this phenomenon are frequently misleading. Commonly an image is drawn of sand dunes reaching out to engulf productive land. But rather than thinking of deserts relentlessly pushing outward, it is more accurate to say that deserts are being relentlessly *pulled* outward by man. And even more significant, in huge areas far away from the actual "edge" of sandy deserts, desiccated, desert-like lands are being *created*, a process that has come to be known as desertification.

Where and at what pace are these trends at work? Documentation is spotty even on current soil conditions in the affected areas, let alone on conditions over a time period long enough to permit precise scientific conclusions. Still, scientists have made valuable estimates of desertification in various regions based on available facts. Perhaps the boldest, if admittedly imprecise, effort has been made by an eminent Egyptian ecologist, Professor M. Kassas. Detailed surveys of climatic data, he observes in a report to the United Nations Environment Program (UNEP), indicate that 36.3 percent of the earth's surface is extremely arid, arid, or semi-arid—categories he combines under the general heading of deserts. Yet a world survey of land conditions based on soil and vegetation data indicates that 43 percent of the earth's surface falls within these categories. The difference of 6.7 percent, Kassas suggests, "is accounted for by the estimated extent of man-made deserts."[2] A collective area larger than Brazil, with rainfall above the level classified as semi-arid, has been degraded to desert-like conditions through deforestation, overgrazing, burning, and injudicious farming practices. These lands support less life of any kind than they would if not misused. And, it should be stressed, this estimate does not take into account the far greater degradation *within* the zones that are arid or semi-arid in the climatic sense.

As subject to error as an estimate like this may be, it is a meaningful reflection of the problem's scale. Regional estimates of the spread of deserts and the degradation of semi-arid lands support the conclusion that desertification is a major world problem and is growing in magnitude. It is a malignancy undermining the food-producing capacity of Africa, Asia, and Latin America.

The southward encroachment of the Sahara Desert is legend, but it is also fact. If the wilder visions of some nineteenth- and early twentieth-century commentators who believed the desert to be engulfing lands at a terrifying, inexorable rate have proven overdrawn, there is little doubt that the desert is gradually being pulled southward. Researchers for the U.S. Agency for International Development estimate that 250,000 square miles (about 650,000 square kilometers) of land suitable for agriculture or intensive grazing have been forfeited to the Sahara over the past fifty years along its southern edge.[3]

Desert encroachment in West Africa has received the greatest international attention recently, but the countries to the east—Sudan, Ethiopia, and Somalia—all face similar threats. Sudan, it is frequently observed, has the potential to become the granary of the Arab world, given its extensive unused lands and unexploited rights to the Nile River waters. If the deterioration of Sudan's natural resources is not halted soon, however, at least some of this potential will dry up. "Perhaps the most striking environmental phenomenon in the Sudan at present," writes Samir I. Ghabbour, an Egyptian zoologist, "is the gradual shifting of vegetational zones toward the south, with an ever-increasing loss of forest and widening of the desert. Desert creeps into steppe, and while steppe loses ground to the desert it creeps into the neighboring savannah which, in turn, creeps into the forest."[4]

The *Acacia* tree, ubiquitous in many arid zones and useful for firewood and forage, was common around the Sudanian capital of

Khartoum as recently as 1955; by 1972, the nearest such trees were ninty kilometers south of the city, according to M. Kassas.[5] Both Kassas and Ghabbour conclude that a combination of overgrazing, accelerated erosion, and the indiscriminate use of fire is responsible for this costly biological march.

The situation in Somalia, as described by Kai Curry-Lindahl of UNEP, is also acute. Although some parts of Somalia are natural desert, extensive additional areas have become deserts or semi-deserts within the last century. Curry-Lindahl predicts that without a radical change in land use practices, inside a few decades the whole country will be desert-like except for some perennial river valleys and the moist southernmost region. He describes the country as "on the verge of a disaster."[6]

While the southward movement of the Sahara has been generating headlines in the early seventies, the desert is also creeping northward toward the Mediterranean. The population of arid North Africa has multiplied sixfold since the beginning of the century, and the destruction of vegetation in Morocco, Algeria, Tunisia, and Libya has accelerated in this period, particularly since about 1930, when the population of these countries began to climb steeply. Intense overgrazing, the extension of unsustainable grain farming, and firewood gathering have all contributed to a deterioration of the agricultural environment. The result, calculates H. N. Le Houérou of the UN Food and Agriculture Organization, is the loss of more than a hundred thousand hectares of land to the desert each year.[7]

With increasingly eroded lands in the Atlas Mountains to the north, and an encroaching desert to the south, food production has stagnated in many areas of North Africa. This one-time granary of the Roman Empire is a chronic, major food-importing region. Soaring proceeds from petroleum and phosphate exports, and remitted earnings of the millions who have migrated to Europe for work, help mask the rural deterioration that is prevalent throughout much of North Africa.

Desertification is by no means limited to the Saharan fringes. It is a major problem in parts of southern Africa, particularly Botswana. Vast semi-arid grasslands in Kenya and Tanzania have been seriously damaged by overgrazing. Barren, desert-like environments have been created by centuries of overgrazing and wood gathering over huge areas of the Middle East from the Mediterranean coast of Israel all the way to Afghanistan—areas that were once at least moderately vegetated. Some such lands would recover if the constant pressure of overgrazing, which is the norm in virtually every Middle Eastern country, were removed. But others have been permanently down-graded by erosion and, in more extreme cases, dune formation. In North America, some experts believe, much of both the Sonoran Desert in Arizona and the Chihuahuan Desert in New Mexico has been created by overgrazing in the few hundred years since the European invasion.[8]

Desert-like lands are also being created in the Argentinian states of La Rioja, San Luis, and La Pampa, and southward encroachment by the Atacama Desert plagues northern Chile. A decade-long drought there in the sixties was accompanied by desert advancement across an 80- to 160-kilometer front at a rate of about 1.5 to 3 kilometers per year.[9]

Apart from intensively irrigated regions like the Nile and Indus Valleys, northwestern India is the world's most densely populated arid zone—a distinction that may turn out to be an epitaph. An average of sixty-one people now occupy each square kilometer of India's arid lands, which include the sandy wastes of the Thar Desert of western Rajasthan; a larger inhabited but desolate area surround-ing it that is often loosely called the Rajasthan Desert; and other dry areas farther south and east. This density is but a small fraction of that supported in nearby irrigated valleys, but it is also, as scientists at India's Central Arid Zone Research Institute recently understated, "quite high in view of limited resources."[10]

The practical consequence of this pressure has been the extension of cropping to sub-marginal lands fit only for forestry or range man-agement, helping to make this perhaps the world's dustiest area. Meanwhile, as the land available for grazing shrinks, the number of grazing animals swells—a sure-fire formula for overgrazing, wind erosion, and desertification. The area available exclusively for grazing in western Rajasthan dropped from thirteen million to eleven million hectares between 1951 and 1961, while the population of goats, sheep and cattle jumped from 9.4 million to 14.4 million. The live-stock population has since continued to grow, while during the decade of the sixties the cropped area in western Rajasthan expanded further from 26 percent to 38 percent of the total area, squeezing the grazing area even more.[11]

Under these circumstances it should come as no surprise that agricultural output is low and, in some regions, falling. So long as current land use patterns continue, the livelihood of the tens of millions living in India's arid lands will at best remain at its current dismal level. At worst, and most probably, a prolonged drought at some future point will mercilessly rebalance people and resources. As it is, relief programs in the arid zones are already chronically sapping the central government's meager funds and food stocks.

Whether, and how fast, the Thar Desert is spreading is disputed. The topographical studies cited in 1952 by the Indian Planning Commission in the then-young country's First Five-Year Plan reached an alarming conclusion: the desert had been spreading at the rate of one-half mile (0.8 kilometers) per year for fifty years, and was annually engulfing thirteen thousand hectares of land. A similar rate of desert encroachment has been widely assumed ever since, but a 1970 study by Indian scientists concluded that expansion by the desert was not now a serious problem. At least part of the discrepancy between this recent conclusion and the continuing assertions by other authorities that the desert is indeed expanding may well be due to definitional differences—what, after all, constitutes the spread of a desert amid a rather desolate landscape? There is no question that thousands of acres of arable land are lost to cultivation each year, and all parties agree that the productivity of an arid area covering more than a fifth of India, an area larger than France, is being seriously impaired.[12] After several decades of accelerating deforestation and chronic overgrazing, much of west and central India is assuming the appearance of a lunar landscape.

Water is the source of desert life, and a long-term shift toward more or less rainfall can, by itself, transform the ecological character of a region. When deserts appear to be spreading, it is natural to question whether a changing climate might be the real culprit. Unfortunately our understanding of climatic change, and factual knowledge of historical weather trends in most desert regions, are too speculative to permit unequivocal conclusions about the role of climate in the current spread of deserts.

Theories of the causes and nature of world climatic change are now almost as numerous as the scientists seriously studying the problem. Major climatic changes over exceptionally long time periods may not be hard to prove. There seems little doubt, for example, that the Indus River civilization flourished four thousand years ago in a moister climate than that region of Pakistan enjoys today. Establishing whether a present-day drought reflects a twenty-year cycle of recurring droughts, a two-hundred-year cycle of oscillating rainfall patterns, the beginnings of a new climatic age, or simply a random event is far more difficult. A review of current theories and knowledge suggests the urgency of research efforts on climatic trends and their causes. It also reveals the undeniable role of people and their livestock in downgrading the carrying capacity of arid lands, and in creating new desert regions.

Some recent theories of climatic change suggest an intriguing, if unproven, compromise between those blaming the climate and those blaming people for encroaching deserts. These theories suggest that the level of rainfall in some important arid zones may indeed be declining, but that this drop has been partly induced by human mismanagement of the land. The increased atmospheric dust produced by overgrazing, rangeland burning, and overcropping, it is surmised, can reduce local rainfall or perhaps even encourage global climatic shifts. Reid Bryson and David Baerreis of the University of Wisconsin, for example, argue that the dense pall of dust annually covering northwestern India and eastern Pakistan has measurably reduced the area's precipitation.[13]

Norman MacLeod of the American University, in Washington, D.C., suggests that the rising amount of atmospheric dust over West Africa resulting from accelerating devegetation of the land is altering climatic patterns and reducing rainfall in the Sahelian zone. Bryson and others have recently postulated *global* climatic shifts, accentuated by human-caused atmospheric dust, which are pushing the monsoon belt of Central Africa and South Asia southward, and thus increasing the incidence and severity of droughts in the Sahelian zone and northern India.[14]

For every theorist of a particular type of climatic change, there are other experts who dispute both the causes and directions of current climatic trends. The more elaborate theories concerning possible shifts in the Northern Hemisphere's monsoon belt are based mainly on extrapolations from data gathered in other parts of the world; the slim local climatic records available in such major desert fringe areas as North Africa, the Sahel, and northwest India do not prove to many the case for climatically induced desert encroachment. While a climatic change in this direction is quite possible, research on weather trends in each of these regions also underscores the heavy responsibility of people in creating new desert lands.

Extensive reviews of available rainfall records in the countries of North Africa reveal no evidence of a decline in rainfall over the last one hundred years, though the desert continues its northward crawl. Indian scientists studying the Thar Desert found no signs of recently increased aridity on its fringes. Dr. E. G. Davy, who carried out a special study of the Sahelian drought for the World Meteorological Organization, concluded that "no serious analysis of available data is known to show a falling trend of rainfall in the zone over the periods for which records are available." The length of the Sahelian drought proves nothing by itself, for comparable droughts have existed in the past and will likely visit the region once again whether or not a major climatic change is taking place. The Sahelian rainfall in the period from 1907 to 1915 was probably just as weak as it was in the recent drought, though a smaller area may have been affected. Reid Bryson, however, maintains that these observed rainfall patterns of the last century strongly support his theory of change. Bryson points out that the resemblance between recent Sahelian rainfall patterns and those of the 1907–15 period is consistent with the postulated reversal of the atmospheric circulation patterns of the intervening period. And while the decade average of total rainfall in the Thar Desert may not have changed, the frequency of severe *drought* years, a calculation lost in decade averages, appears to be increasing.[15]

Whatever the results of climatic research, an inescapable central lesson to be drawn from the experience of the world's drier zones— whether in the American Great Plains or the Sahel—is that droughts are an unavoidable aspect of the arid environment. Though they cannot be predicted with any precision, they should never come as a shock. Over and over again droughts are perceived as unexpected natural disasters just like tornadoes or earthquakes; the real calamity arises from the failure of societies to mold their habits to fit environmental reality. Human cultural patterns in the desert must be re-shaped to survive the driest years, not to push the land to its limits in years of favorable rainfall. Any other approach promises death and dislocation every time the rains fail for long. And as the number of people and animals living in the arid zones climbs, while the quality of the land on which they must live simultaneously shrinks, the impact is bound to be more catastrophic with each successive drought.

Just as human suffering in the desert peaks during a drought period, so do the destruction of vegetative cover, wind erosion on cultivated fields, and the formation of unusable wastelands. Several consecutive years, or even decades, of reasonably good rainfall encourage the growth of forage and thus of herds, and the extension of cultivation to lands more safely left in grass. False confidence about the carrying capacity of the land often grows among farmers and governments—a recurring phenomenon in both the United States and Africa. Then, when the rains finally fail, overabundant animals eat every available blade of grass on the weather-decimated rangeland until finally many starve to death, especially around wells, which attract a heavy concentration of livestock. Goats start climbing trees to eat the last bit of greenery, leaving behind a wooden skeleton for the firewood gatherers. Crops fail to take root in the parched ground; the bare plowed soil yields to the wind; and a dust bowl is created. Sand dunes appear where none existed before, and, if the rains stay long absent, a new patch of desert may appear. Nomads, farmers and their surviving herds retreat before denuded lands, setting in process a self-reinforcing negative spiral as successive areas become over-crowded and overgrazed due to the influx of refugees.

The years immediately following a severe drought can again bring a bogus optimism about the land's capacity. By killing off a share of the herds and driving some of the people from the region, nature has temporarily restored an equilibrium of sorts. Decimated herds put only a small fraction of the earlier pressure on grazing lands, and, with better rainfall, grasses and shrubs spring back to life wherever the soil has not been seriously damaged. Refugees trickle back and, unless some new force has intervened to break the pattern, the same deadly cycle begins again. To the extent that desertification has occurred, though, the area available for grazing and cropping is reduced, even while the spread of modern medicines, livestock-disease control, and well construction encourage ever larger populations of humans and livestock. Thus without a concomitant transformation of farming and pastoral techniques, "development" activities like well construction or cattle inoculation can have a negative net impact on human welfare along the desert fringes.

While there is great disagreement about the causes and directions of any current global climatic changes, a number of the world's leading meteorologists predict a period of greater global climatic instability in coming decades than that experienced over the first sixty years of this century. The slight global cooling trend in evidence since 1940 may, for complex reasons, mean a more variable climate throughout the world.[16] If this assumption is accurate, the consequences for agriculture almost everywhere will be negative, since crops and farming habits are closely geared to a predictable climate.

Likely to be hardest hit by such instability would be the dry zones, such as the Saharan margins and northwestern India, which would face more frequent droughts and monsoon failures than is customary. The familiar arid-zone cycle of expanding herds, followed by drought and economic and ecological collapse, is being recapitulated on a century-long scale if the theories of a new climatic instability are correct. The climate of the last several decades, it is posited, has been unusually benign and stable compared with that throughout most of the earth's history. These have also been the decades when traditional natural checks on human and livestock populations everywhere were to a great extent overcome by modern science. If some arid zones are entering an era of more frequent drought, their capacity to support life will be generally lower in the future than in the recent past, and the already acute man-made ecological pressures on them will be intensified.

The possibility of climatic change clearly complicates prospects in such dry zones as the Sahel, but it does not alter the basic strategies for reversing ecological deterioration—the outlines of which have long been known. In virtually every case the two keys to recovery of the land are a sizable reduction in the number of grazing animals, and a halt to the myopic spread of cultivation every time a few years of good rainfall lures desperate farmers onto marginal land.

Making such a statement is, of course, easier than implementing it. Decades of cumulative environmental deterioration, despite the Cassandra-like warnings of scientists, attest to the seemingly implacable nature of the problem. The basic dilemma is that what is essential to the survival of the society often flies in the face of what is essential to the survival of the individual. The individual nomadic family, largely isolated from the cash economy of the cities, needs many dozens of animals just to meet its basic needs for milk and milk products—often the principal foods. A surplus of livestock beyond the basic minimum is both an investment and a form of insurance for the drought years, when some animals can usually be sold if the group moves toward the cities, and when some are bound to die in any case. The individual farmer sees little choice but to plow up high-risk sub-marginal fields. On the best fields, yields may be too low to provide enough food for the local populace, and may even be falling as the traditional fallowing custom is abandoned to compensate for the shortage of good new farmlands. Plantations of cash crops for export set up by the government, a few wealthy farmers or foreigners may have squeezed the subsistence farmers out of important cultivated areas. Moreover, as long as children are a form of insurance against inevitable family deaths, and are valuable contributors to the labor pool for jobs like tending herds, couples in the arid zones will naturally place a premium on large families.

A better future for the arid lands and their peoples, then, depends upon a fundamental shift from a system in which the exercise of personal aspirations encourages social suicide to an institutional structure in which those working to better their own lot are also furthering the long-term welfare of society. It will have to be a system in which livestock is valued for quality and economic value rather than simple numbers; in which farmers have the knowledge and equipment to grow enough food on the best-suited lands without running down the land's long-term fertility; and in which families see an advantage in remaining small. A successful new order along these lines will almost certainly involve the closer economic integration of the arid zones with the cities and regions with more moderate climates. An inward flow of resources, information, and goods to the desert edges is essential to a new order there; equally essential, in turn, is the quite feasible outward flow of high-valued meat and other agricultural products.

Faced with the urgency of such a radical transformation of life on the desert margins, some analysts retreat into oversimplified solutions that do not address the fundamental predicament. As nomads find their movements restricted and their grasslands deteriorating, some governments see a need to settle them down at almost any cost. Watching the ubiquitous goat destroy trees, shrubs, and grasses, a few ecologists advocate the total elimination from arid lands of this hardy, well-adapted animal. As water becomes scarce, local leaders demand that national governments or international aid agencies dig wells, without simultaneously instituting controls on the size of local herds and on their access to the pastures surrounding a new well.

In the desert, as elsewhere, planners have much to learn from the plants, animals and cultures that have survived over the centuries under extraordinarily adverse environmental conditions. If the ecological balance historically maintained by most nomadic groups was a rather wretched one, one predicated on high human death rates, it also made remarkably resourceful use of the life-defying desert. In popular mythology, nomads are often pictured as aimless wanderers, but in fact, nomadic movements are nearly always systematically attuned to the seasonal rhythm of climate and plant life; they are geared to provide adequate forage for the herds throughout the year and to permit the regrowth of grazing lands.[17]

A return to an earlier historical age is probably no more desirable than it is possible. The harsh system of unmitigated natural selection that underlay the past viability of nomadic systems is not a condition ethically acceptable to most today. In any case, modern medicine has trickled into the arid zones well ahead of such other modern accouterments as advanced agricultural technology, thereby ensuring lower death rates than those of the past. National boundaries now impose artificial divisions on natural ecological zones and impair the traditional movements of nomadic groups, while the spread of sedentary agriculture further constricts their migrations. It also appears likely that some nomadic groups never achieved an ecological equilibrium with their habitat. The Masai of East Africa, for example, seem to have gradually shifted southward over many centuries preceding the colonial era, abandoning denuded and overgrazed lands.[18]

Yet some modernized version of the nomadic way of life may be the only possible means of exploiting the protein-producing potential of the more arid desert fringes, and the only way to save the livelihood of many of the millions inhabiting these zones. Huge regional management schemes, in which clan leaders regulate grazing and migratory movements according to natural conditions and the advice of range specialists, are one possibility now being explored in Niger. A system like this would ideally retain much of the flexibility of nomadic ways, while also permitting technological progress. Once minimal control of grazing is established, wells and pasture improvements can be introduced with less chance that their long-run impact will be twisted.

A basic prerequisite for the success of any livestock scheme is a major reduction in herd sizes. According to some specialists, the number of grazing animals maintained in the Sahel up to 1972 and 1973, as the lengthy drought reached its climax, was at least double what the zone's ranges can sustain without damage. The drought cut animal numbers steeply, but not enough to put grazing and grasses back into balance. Yet numbers deceive: if the total number of animals were maintained at *half* the 1971 level, and simple management techniques were implemented, the region's output of meat and milk could easily be *double* its former level.[19]

This somewhat paradoxical formula arises from the nature of livestock growth. Of all the food consumed by a grazing animal, roughly the first half is required just for physiological maintenance; the next fourth is largely required for reproduction, and the final fourth goes into milk production, growth, and fat storage. Any cutback in feeding forced by overgrazed, depleted pastures is mostly at the expense of these final functions. Of course, the benefits of a more efficient grazing system that emphasizes productivity over herd numbers can be realized only when a marketing system to reach consumers with the valuable livestock products is also created.

Improved farming in the sedentary zones is just as crucial as controlled grazing; some would give it even higher priority. In past decades most research and investment in arid zone agriculture concentrated on cash crops like cotton and peanuts for export, and on large-scale irrigation schemes that hoped to bring desert regions under intensive production of food or fiber. The simple subsistence farmer, growing millet or sorghum for his or her family and, in good years, for trade with nomads or urbanites, has frequently been neglected, with sorry consequences for the land. Cash crops, the principal source of foreign exchange for many arid countries, can be one key to economic progress. But if their expanded cultivation is not accompanied by careful land use planning, taking into account local food and land needs, and if a major share of the income they produce is not earmarked for the betterment of rural economic and social prospects, the main impact may be to worsen the lot of the rural poor. Satellite photographs indicate that, due to the combined impact of rapid population growth, traditional technologies, and the extension of cash cropping, Sahelian farmlands once left fallow for ten to fifteen years to allow a regeneration of fertility and moisture are frequently now idled for three years or even less.[20] Under these circumstances, the land is sapped of its natural protection against drought.

Neither population pressures nor claims for other land uses will ever permit a return to the former fallow system. The only alternative is to move toward new cropping systems that minimize erosion by maximizing vegetational cover on the land, and to preserve fertility through crop rotations, organic fertilizers, scientific fallowing, and, perhaps, the introduction of chemical fertilizers. Agricultural techniques that prevent dust bowl conditions like those that emerged in cropped areas of West Africa in 1972 and are fairly chronic in northwestern India have been developed and proved effective in such countries as Australia, the Soviet Union, and the United States.

In addition to the adoption of new agricultural methods, tree-planting programs are urgently needed throughout virtually every region of dryland agriculture in Africa, Asia, and Latin America. In the American Great Plains, many thousands of windbreaks have been planted since the 1930s and have, along with other improved practices, helped stabilize a system that once threatened to become a permanent dust bowl. Reforestation programs will also help relieve the critical shortage of fuel for cooking that now plagues every arid zone in the developing world.

Tree-planting programs directly on the fringe of sandy deserts can also help stem the tide of encroaching dunes. While localized programs of dune stabilization have been tried in most of the threatened countries, Algeria is setting an example for the rest of the world with its bold new plan to plant a sixteen-kilometer-wide forest barrier all the way across its fifteen-hundred-kilometer expanse. The Algerian national youth corps and army have been given the job of reforestation; through this program they may do more to safeguard the country's "national security" than they could through any other undertaking.

The recent lengthy African drought and famine will stand out as a major event in the history of arid lands. Not only the drought's unusual severity, and its bitter toll of lives, but also its timing account for this; it arrived in an era of global communications and rising awareness of ecology.

At the intellectual level, at least, the tragedy of the Sahel has had a catalytic impact comparable to that of the American dust bowl of the 1930s. For several decades ecologists and range specialists have been warning, to little avail, about the growing pressures being placed on delicate arid environments like the Sahel, but their perception did not penetrate very deeply into the priorities and working programs of governments and aid agencies. The understanding of the extent and causes of desertification, however incomplete, is by now far too clear for any responsible leader to ignore. A 1977 United Nations Conference on Desertification will further deepen both the technical knowledge and the political awareness of this global challenge. Even so, solutions will not come easily or quickly in the Sahel or any other distressed arid zone, for the social and economic obstacles to a new pattern of existence are formidable.

5. Refugees From Shangri-La: Deteriorating Mountain Environments

A<small>N</small> <small>UNUSUAL MEETING</small> was convened in Munich, Germany, in December, 1974. Any organizing principle, any common thread among the participants, would have eluded an outsider. The group included biologists, anthropologists, foresters, ecologists, economists, geographers, businessmen, and civil servants, and they had traveled to Munich from Europe, North and South America, Africa, and Asia. What drew this diverse group together was a shared concern for a problem seldom recognized as deserving attention in its own right: the deterioration of mountain environments in the poor countries.

Mountains, in contrast to deserts—another sparsely populated, economically marginal portion of the earth—have been largely neglected by researchers and governments. Perhaps the fact that the earliest human civilizations evolved in arid regions has something to do with the special fascination deserts have always held for humankind. Whatever the reason, a major international program of arid-zone research was initiated by the United Nations as early as 1951, whereas coordinated interdisciplinary research on mountain environments is just now getting underway.

Highlands occupy about one-fourth of the earth's land surface but provide a home for only a tenth of the world's people.[1] Still, it is curious that mountain ecosystems have been ignored so long in comparison to other natural areas, for history has repeatedly shown that when ecological changes take place in the highlands, changes soon follow in the valleys and the plains. And while only 10 percent of the human population lives in the highlands, another *40 percent* lives in the adjacent lowland areas, their future intimately bound to developments on the slopes and plateaus above.

For all the diversity that characterizes their land and peoples, the three major mountain ranges of the developing regions—the Himalayas, the Andes, and the East African highlands—present a rather uniform set of environmental and economic challenges. What strikes the casual observer of ranges like these is their stark immutability, their massive grandeur, but in fact they are among the most fragile ecosystems on earth. Steep mountain slopes can seldom sustain the degree of cropping, woodcutting, and grazing that is customary in flatter areas. Yet all three practices are escalating in order to cope with rising populations throughout the mountains. And the inherent difficulty of adapting cultural practices to rapidly changing environmental circumstances is exacerbated by the fact that, in many countries, mountain populations are those with the least income, the least education, and the least political power. Basic development activities like providing farmers with technical advice, children with schooling, or parents with family planning services are severely hampered by the craggy geography of high mountains.

When the environment starts to deteriorate on steep mountain slopes, it deteriorates far more quickly than on gentler slopes and on plains. And the damage is far more likely to be irreversible. The mountain regions are poor not only in economic terms; many areas are rapidly losing any chance of ever prospering as their thin natural resource base is washed away. Degenerating economic and ecological conditions in the mountains, in turn, often push waves of migrants into the lowlands, leaving behind an aged, dispirited population incapable of turning around the negative trend. Mountain people have little choice but to follow their soils down the slopes. This movement simultaneously adds new pressures to the lowland ecosystems, which are often under stress themselves, and to the generally overburdened cities.

The net result of current trends, as evaluated by the international committee of experts recently established by UNESCO to study mountain environments, is "accelerating damage to the basic life-support systems" today in practically every mountainous region of Asia, Africa, and Latin America. "Within the last decade there has been a marked increase in the destructive clearance of forests, in flood damage and silting, in soil erosion and the explosive spread of pests. . . . In sum, human pressures on tropical high mountain ecosystems are increasing nearly everywhere. . . . They are unusually prone to sudden, rapid, and irreversible loss of soils if slope stability and vegetation cover are disturbed."[2]

The unanimity of the concern of scientists and laymen with experience in the mountains of every continent was striking at the 1974 Munich gathering, which was called by individuals who hope to institutionalize greater cooperation in research and public education on mountain problems. The group warned of "the irretrievable loss to human use of the developing world's mountain resources—in some cases within one or two decades—unless the present rate of deforestation and land mismanagement can be halted. . . . The ever-widening circle of destruction originating in the mountains is spreading to the plains, river systems and harbors."[3]

The mountain regions often possess great economic potential as sources of hydroelectric power, of valuable timber, minerals, pastures, orchards, and scenic natural refuges that are in growing demand and diminishing supply. But a distinction must be made between the theoretical potential and the present-day realities. For without a massive effort to preserve and restore the ecological integrity of the mountains, within a few decades idyllic panoramas will become barren eyesores that perennially overwhelm the lowlands with devastating torrents and suffocating loads of silt.

There is no better place to begin an examination of deteriorating mountain environments than Nepal. In probably no other mountain country are the forces of ecological degradation building so rapidly and visibly. This kingdom of twelve million people is minuscule by Asian standards, but it forms the nucleus of one of the world's strategic ecological nerve-centers. The Himalayan arc, stretching from Afghanistan through Pakistan, India, Nepal, and Bhutan to Burma, forms an ecological Gibraltar whose fate will affect the well-being of hundreds of millions. From the Himalayas flow the major rivers of the Indian subcontinent—the Indus, the Ganges, and the Brahmaputra—which annually bring life, and sometimes death, to Pakistan, India, and Bangladesh.

Nepal itself does justice to irrepressible superlatives. It boasts the world's highest mountain, and its features, among the most varied of any country, range from the glaciers of Everest to warm tropical forests on its southern fringe. But in this land of unexcelled natural beauty live some of the world's most desperately poor. Nepal also faces what could be the world's most acute national soil erosion problem.

Popular image and reality rarely diverge so widely as they do in the case of Nepal. Most know the kingdom as the photogenic home of Mount Everest, as an exotic Shangri-la sprinkled with pagodas and quaint villages tucked away in the folds of the Himalayas. The colorful, anachronistic coronation of its king in early 1975 sparked a flurry of articles on the land's charms in the international press, and pictures of the visiting socialites overshadowed those of the king and queen themselves on society pages around the world.

The façade of romance and beauty remains intact, but behind it are the makings of a great human tragedy. Population growth in the context of a traditional agrarian technology is forcing farmers onto ever steeper slopes, slopes unfit for sustained farming even with the astonishingly elaborate terracing practiced there. Meanwhile, villagers must roam farther and farther from their homes to gather fodder and firewood, thus surrounding most villages with a widening circle of denuded hillsides. Ground-holding trees are disappearing fast among the geologically young, jagged foothills of the Himalayas, which are among the most easily erodable anywhere. Landslides that destroy lives, homes, and crops occur more and more frequently throughout the Nepalese hills.[4]

The decade of the 1960s was a period of record economic growth for most of the world's poor countries. For the seven and a half million Nepalese who live in its hilly region, however, as well as many of the tens of millions of hill residents of northern India, the decade was characterized by what World Bank analysts and key Nepalese planners have both frankly labeled a "deteriorating economy." Per capita income, already well below one hundred dollars a year, is falling, as is per capita food production. Undernourishment is widespread and combines with the near total lack of modern medical services to ensure one of the highest infant mortality rates in the world. As many as three hundred out of every thousand babies born in Nepal's hills die before their first birthday.[5]

Topsoil washing down into India and Bangladesh is now Nepal's most precious export, but one for which it receives no compensation. As fertile soils slip away, the productive capacity of the hills declines, even while the demand for food grows inexorably. Some terraces are expertly managed and ecologically stable, others continue to be farmed despite their waning productivity, and others reach the point of no return and are abandoned. In the country's most densely populated region, the eastern hills, as much as 38 percent of the total land area consists of abandoned fields.[6] Once these slopes are left to face the violent monsoon downpours without protective vegetation, their more fertile soils may be lost forever and their potential usefulness to people permanently reduced.

While acceleration of naturally heavy erosion is the chief threat, the declining fertility of the hills stems in part from another problem as well. Nepalese farmers have always assiduously applied the available animal manures to their fields as fertilizer, but in some regions the fields now receive less manure than in the past—well below the full amount necessary to preserve high fertility.[7] This is partly because herds have not grown as rapidly as the cultivated area; the hills are already overgrazed and fodder of any kind, whether tree leaves or forage crops, is scarce. Even more ominously, farmers faced with an unduly long trek to gather firewood for cooking and warmth have seen no alternative to the self-defeating practice of burning dung— sorely needed for fertilizer—for fuel.

The average hectare of arable land in Nepal's hills must now support at least nine people. This is a person-to-land ratio comparable to that in Bangladesh or Java, where far more fertile soils and more developed irrigation systems often permit several crops a year on the same land. And in the hills, the Nepalese government realizes, "there is absolutely no scope whatsoever for bringing new land under agriculture."[8]

If Nepal's borders ended at the base of the Himalayan foothills, the country would by now be in the throes of a total economic and ecological collapse. Luckily, the borders extend farther south to include a strip of relatively unexploited plains known as the Terai, an extension of the vast Indo-Gangetic Plain of northern India, one of the world's most productive agricultural areas. The Terai suffers seasonal floods and is heavily vegetated, and until mid-century the high incidence of malaria precluded heavy settlement. Once an effective malaria eradication program got underway in the fifties, however, a rush to colonize the region was inevitable, for the Terai was bounded on the north by increasingly cramped hills, which offered declining economic prospects, and to the south by the badly overcrowded plains of India.

The Nepalese government recognizes that the controlled settlement of the Terai is an essential step toward taking some of the pressure off the hills, but it lacks the power to keep the land rush under any meaningful control. In the decade from 1964 to 1974, 77,700 hectares of Terai forestland were officially distributed by the government to settlers. But more than three times that amount was cleared illegally by migrants from the hills or, perhaps even more significantly, from India. Though the rate of population growth for Nepal as a whole is about 2.3 percent, migration joins the natural increase in the Terai to push its growth rate up to 3.6 percent. Emigration does simultaneously hold down the rate of growth in the hills to an average of 1.3 percent, though it remains much higher in some hill regions.[9]

The presence of undeveloped arable land in the Terai provides Nepal with some breathing space in which to reverse the downward spiral of population growth, land destruction, and declining productivity in which it is now caught. Yet the length of this reprieve is frequently exaggerated, both by outsiders and by some Nepalese. Analysis of satellite photographs indicates that less than half the remaining three-quarters of a million hectares of Terai forestlands will be suitable for cultivation.[10] Much of the remainder is too steep or swampy for economical cultivation, or must be kept under forest for watershed protection. If migration down into the Terai continues at the pace of the last ten years, all the good farmland will be occupied in little more than a decade.

The reckless settlement of the Terai is already taking its toll on the area. Though the soils tend to be of good quality, as a result of primitive technology and the doubling of the regional population every twenty years, the Terai's food surpluses, which help keep afloat the national balance of payments, are gradually falling. Indeed, if recent trends continue, Nepal, which has long been a net exporter of food to India thanks to its Terai surplus, may well join the ranks of food-importing nations by the late seventies—if it can find the foreign exchange to do so.

Agricultural modernization—better seeds and innovative techniques, land reform, extension and marketing services, and so on—will quickly have to replace the extensive spread of farming to new lands if Nepal is to avoid an acute food crisis. Yields must be raised dramatically on the best farmlands, so that the dangerous spread of cultivation to marginal lands can be reversed. The potential for raising yields may be greatest in the Terai, but the hills cannot be neglected any longer either. Migration is a temporary palliative, not a long-term solution; a large share of the country's citizens will always live in the hills regardless of the quality of life they provide, and soon there will be no place else for the people to go in any case. Furthermore, the devastation in the hills is exacting a heavy, if as yet unmeasured, price on the potential productivity of the Terai. Those living near their banks tell government officials that the incidence of flooding by swollen rivers coming down from the mountains is increasing. Furthermore, soil conservation officials observe that the bed level of many Terai rivers is rising from six inches to one foot every year.[11] Not only does this guarantee wider floods from even normal volumes of water in the monsoon season; it also causes the river courses to meander about, often destroying prime farmland as they go.

The cultures of Nepal have not had to contend with such a serious problem of land scarcity before. While terracing has always been necessary to farm the hills, the development of a national conservation ethic has never before been so essential to their survival. A review of successive reports filed in Kathmandu, the country's capital, by Western forest-conservation directors in the fifties and sixties makes fascinating reading; these were, quite literally, plaintive cries in the wilderness. As early as 1954, a UN-supplied forester wrote that deforestation in the hills was "becoming catastrophic and erosion is causing the loss not only of property but also of human lives." One of his successors followed up in 1967 with these words, after an apparently frustrating five years in the country: "It seems an extraordinary thing that in a poor mountainous country such as Nepal, where every foot of soil is precious and is required to produce the necessities of life in the shape of food and shelter, the brown soil-laden rivers should go unnoticed during the monsoon, and the fact that they are carrying away forever the basis of the very life of the people should mean nothing at all to the vast majority."[12]

By now, however, the scale of destruction is hard to ignore, and this, in combination with the continuing efforts of dedicated individuals both inside and outside of the Nepalese government, is stirring new interest in the integrity of the environment. The country's influential National Planning Commission recently expressed its concern with an urgency unsurpassed by any party. Soil erosion, the commission fears, is "almost to the point of no return. . . . It is apparent that the continuation of present trends may lead to the development of a semi-desert type of ecology in the hilly regions."[13]

Translating official awareness into meaningful programs on the ground is no mean task, particularly in a country with such limited resources and unique transportation and communication problems. In August, 1974, a Department of Soil and Water Conservation was finally formed within the Ministry of Forests. It hopes to establish demonstration projects in villages scattered about the country, but in early 1975 the Department included just sixty-seven employees, of which fewer than a third had professional training. According to national development policy, this department is slated to expand to 167 employees over the next five years—a mere beginning.

Faced with the inevitability of absorbing many of the consequences of ecological degradation in Nepal, Indian officials have encouraged the Nepalese to attack their problems head-on and have even provided limited assistance in land-management research. When mounting silt loads threatened the viability of an Indian diversion dam and irrigation project on the Kosi River, built just where the river emerges from Nepal, India in 1969 sponsored a research mission into the Nepalese portion of the Kosi watershed, which also extends north of Nepal, deep into Tibet. The team determined that while heavy siltation was inevitable given the geology of the catchment, a significant share of the sediment was produced by improper cultivation and overgrazing on the steep interior hills. Unfortunately the group was unable to examine the extent of land deterioration in the Tibetan portion of the watershed, which is nearly as large as Nepal's. The group warned that the Kosi catchment, which includes most of eastern Nepal, was now "one of the worst eroded in the world" and that "the hill slopes, which are generally cultivated along the gradient, quickly lose their topsoil and the ability to grow crops or sustain any economic type of vegetation."[14]

The Indians are worried about environmental trends in Nepal, and with good reason, but the fact is that virtually identical problems plague even larger hilly expanses within India itself in such states as Himachal Pradesh, Uttar Pradesh, Assam, and Jammu and Kashmir. In large mountain regions, the fertile valley floors have long been overcrowded, and cultivation is constantly pushed onto steeper slopes by population growth in the absence of non-agricultural employment opportunities. On millions of hectares there is no longer any topsoil at all, just a rocky sub-stratum lacking organic matter or fertility. Forests are receding under the combined pressures of shifting cultivators; uncontrolled herds of goats, sheep, and cattle; and wood gathering for home consumption or sale.[15] In many areas, firewood is so scarce that dung must be burned for fuel.

Many farm families cannot subsist on the output of their small, often infertile landholdings, which in many cases are fragmented into widely separated plots the size of table-tops. As a result, a high proportion of the able-bodied men migrate to the plains to find seasonal work, usually returning to help with the planting and harvesting. Families see little choice but to squeeze from the land what benefits they can, regardless of any possible long-term consequences for its fertility or for the farmers downstream. Even so, the battle for subsistence through agriculture is often lost, and the death rate among infants and children is high.

Moving westward in the Himalayas, into northern Pakistan and then Afghanistan, the outlook for the mountain environment is hardly more encouraging. Both countries are mainly arid and desert-like, which means that the limited available forests in the mountains must bear an especially heavy burden. Pakistan today, for example, classifies only 3.4 percent of its land as forest, and nearly all of that is concentrated in its hilly northern provinces. Large stretches in these hills have been visibly deforested within the last century.[16] On top of the growing pressures from agriculture and overgrazing, the remaining forests must satisfy the burgeoning national wood need for construction, industry, furniture, and fuel. While farmers lop the branches off of trees to provide their animals with fodder and their homes with fuel, timber concessionaires respond to some of the world's highest lumber prices in the cities below by clearing large stands of timber, often heedless of the environmental consequences.

Through these northern hills, after passing through India, flows Pakistan's jugular, the Indus River system. The water of the Indus and its six major tributaries is about all that stands between a bustling, densely populated civilization and a deserted, sandy wasteland. With erosion rampant in the uplands, the exceptionally heavy silt load carried by these rivers is rendering the country's expensive new reservoirs useless with startling rapidity, and has become a pet subject of editorials and political speeches over the last decade. Pakistani foresters are pressing ahead with an excellent series of forest-resource and land-management surveys in the mountain regions, but obtaining the funds and political commitment needed to act on their research findings is proving more difficult. The absence of any type of regional cooperation between Pakistan and India—whether for academic research on their shared watersheds or action programs to halt their degradation—further hampers the task of ecological protection. In Pakistan, as elsewhere, the sheer scale of the problem is overwhelming the scattered attempts to reverse the negative trends.

The barren, infertile landscape presented by most of Afghanistan, and by the Zagros and Elburz mountains farther west in Iran, are fair warning of what lies ahead for parts of Pakistan, India and Nepal if prevailing trends are not reversed. A United Nations team visiting Afghanistan in the mid-1960s found the country's river basins "remarkable for their sparseness of vegetation and the paucity of animal life. The upper catchments are often bare rocky mountains with almost no soil cover and very little vegetation." A German agriculturalist who worked five years in Paktia Province, where most of Afghanistan's few remaining trees are standing, laments their imminent demise by writing: "In Afghanistan the last forest is dying—and with it the basis of life for an entire region."[17] Afghanistan was never blessed with ample rain and fertile soils; even many centuries back the forests covered only a small area. By now, however, they have been reduced

to well below one-tenth of 1 percent of the country. As in neighboring
Pakistan, a profitable lumber trade, browsing animals, and the rising
number of cultivators are conspiring to undermine the mountain
ecology.

Halfway around the world from the Himalayas, dominating the
west coast of South America, stand the Andes Mountains. The long-
est range in the world, these mountains form what looks like the
misplaced spine of an entire continent, stretching all the way from
Venezuela through Colombia, Ecuador, Peru, Bolivia, and Chile
almost to Cape Horn. This massive ridge juts forcibly from a deep
trough in the Pacific, then quickly falls almost to sea level and the
tropical forests where the Amazon's tributaries begin their five-thou-
sand-kilometer course to the Atlantic. The close juxtaposition of arid
coastal deserts, snow-covered peaks, and steamy jungles gives some of
the Andean countries a terrain of even greater contrast than that of
Nepal. In places, a distance of only 320 kilometers separates points
six thousand meters below sea level from others more than six thousand meters above. Travelers in the Andes invariably register surprise
when their train or bus crosses the continental divide near the west
coast, in spots only one hundred kilometers from the Pacific.

The major temperate-zone mountain ranges, such as the Alps and
the Rockies, have always been thinly settled in comparison with the
surrounding lowlands. Just the opposite is true of the Andes and the
Ethiopian highlands, which have been densely populated for many
centuries, and are bordered by sparsely populated tropical rain-forests
or deserts. The highland valleys and plateaus hidden among the steep
Andean peaks still contain a majority of the people in the principal
Andean countries. Though rugged, these uplands present a natural
environment far more hospitable to permanent agriculture and human life than the surrounding areas. They include such regions as the
Cuzco Basin, once the seat of the remarkable empire of the Incas,
and the Bolivian *altiplano*. This treeless tundra-like plain, the homeland of the llama and the alpaca, surrounds the basin of Lake
Titicaca, the world's highest navigable lake.

From Ecuador southward, much of the narrow coast separating
the Andes from the Pacific Ocean is a leafless desert, striped with the
irrigated oases that surround each of the sixty or so rivers flowing
down the westward side of the mountains. Some of these flow only

seasonally, but most flow the year round, and for centuries they have supported intensive agriculture and compact human settlements. Across the mountains to the east, in the jungle of the upper Amazon, an area that encompasses more than half of Peru and Bolivia and parts of Ecuador and Colombia, tropical diseases and the limited producing capacity of tropical agricultural systems have historically kept the population low. Over the last few decades, however, this humid tropical zone has been on the receiving end of one of the great human migrations of the century: the movement of highland people throughout Latin America down into the humid lowlands. The spread of modern medicine to the jungles has paved the way for this procession; its causes, however, lie in the slopes above. Most of its participants, like those in the even larger procession to the region's cities, are refugees from the deteriorating agricultural systems and the exploitative social systems of the mountains.

The pressure of human numbers on the environment is not a new phenomenon in the Andes, nor is knowledge of the farming techniques necessary to save the soil from accelerated erosion. Some Andean valleys doubtlessly provided sustenance for more people five hundred years ago than they do today, and did so at less cost to their long-term fertility than is exacted by present-day farmers. The unusual history of ecological deterioration in the Andes begins long ago, in the millenia preceding the European discovery of the New World.

By about 500 A.D. one of the major cultures of pre-Columbian America was emerging in the coastal valleys and upland plateaus and gorges of Peru. In relatively independent small settlements ruled strongly by priestly and warrior castes, the fruits of many centuries of technological and political development began to crystallize in societies able, like the earlier Egyptians and Sumerians in the Old World, to master the hydraulic cycle and thus create a large, complex civilization. These ancient Peruvians created an intensive agricultural system in the arid valleys of the coast, and on the slopes and plains of the mountains, by performing what historian William H. McNeill has termed "extraordinary feats of water engineering," and building "terraces as elaborate as any the world has ever seen."[18] The terraces held down erosion, while fallowing was practiced to preserve the fertility of the soil.

The modern world owes a great debt to these ancient Andean

cultivators. The area now called Peru was one of the world's great centers of plant domestication, and the sixteenth-century Spanish conquerers of Peru (and Mexico) found many unknown vegetables to enrich the diets of their homelands, vegetables like maize, squash, numerous strains of beans, manoic, the tomato, sweet potato, peanut, and avocado.[19] Most significant of all was the white potato, which had evolved, and remains, as the principal staple of the Andean residents of the higher altitudes, where maize cannot grow. The potato quickly spread through the Old World, contributing to an increase in European population by greatly augmenting food supplies there. Today production of the potato in Europe and the Soviet Union is several times that in the New World, where it originated.

Hydraulic engineering and social regimentation both reached a climax in the fourteenth and fifteenth centuries, when the Incas of the Cuzco Valley overran the city states of the Andes and the coast to establish the empire of Ttahuantinsuyu, the Four Quarters of the World, otherwise known as the Inca Empire. These Indians of the mountains extended their rule over an area the combined size of France, Switzerland, Italy, Belgium, the Netherlands, and Luxembourg. Its span was equal to that of ancient Rome, from Britain to Persia. Considering the rugged terrain into which they forged, which included parts of what are now Colombia, Ecuador, Peru, Bolivia, Argentina, and Chile, the Incas imposed extraordinary political and economic control comparable to that of the Pharaonic societies of Egypt. Their skillful waterworks and wide mountain roads were marvels to the sixteenth-century Spanish conquerers who copied them.

The Inca Empire fell in 1532 under the combined impact of a major succession crisis, quite possibly a devastating epidemic of smallpox introduced by Francisco Pizarro and his followers, and, finally, Pizarro's military invasion. The Incas had successfully created, in a fragile environment, a sustainable agricultural system that minimized the damage to the productivity of the land. Yet, by the latter years of the empire, there is some evidence that the pressure of population on the limited arable area was beginning to show.

Felix Monheim, a German geographer, cites the extensiveness of the costly irrigation, terracing, and artificial field elevation works, as well as the empire's food-reserve policy, as evidence that, by the early sixteenth century, the carrying capacity of the land had been reached or perhaps even, in unfavorable weather years at least, surpassed. Even before this time the central Andes were largely deforested and most mountain residents were dependent on the dung of llamas for cooking fuel.[20]

The extent of incipient land degradation in the Incan empire is open to debate. More certainly, any threat of overpopulation quickly dissipated following the Spanish occupation. Historians disagree about the population of the Inca Empire at its height; estimates range from several million to over sixteen million. But in any case, the combined influence of wars, forced labor in unhealthy silver and mercury mines, and European diseases sharply reduced the population of the Andes—probably by as much as three-fourths in the first century of colonial rule.[21]

While this decimation of the population obviously reduced any immediate pressures on the land, it did contribute, ironically, to the emergence of quite unfavorable conditions for land conservation in the following centuries. A small population and vast open spaces facilitated the adoption of huge European-owned estates, including cattle ranches on the highest plateaus and crop-growing haciendas in the middle and lower valleys. The former Inca system of central rule and labor tributes was replaced by one in which Spaniards favored by the Crown were given large landholdings, and control over the Indian villages they contained. The estate owners extracted produce and labor from their Indian charges, while the government requisitioned further Indian labor for the mines, plantations, and cities. Though these conditions of virtual slavery passed with political reforms and the nineteenth-century emergence of independent nations, by the mid-twentieth century many of the Andean Indians remained bound in servitude to landed proprietors. And outside the large estates on which many have lived and worked are overcrowded, fragmented landholdings unable to provide sustenance for their populations. By 1957, an historian of the Incas could write of the Andean Indians: "Centuries of exploitation, degradation, and neglect have reduced them to a stolid, poverty-stricken people. . . . Despite their greater freedom from regimentation and regulation, their lot is probably less desirable than in Inca days."[22]

Unfortunately, the decimation of the Inca population and social order was accompanied by the loss of the conservation ethics and know-how of the former empire. With scarce labor concentrated in the mines and on the hacienda fields of fertile valley floors, most of the terraces and irrigation facilities constructed over previous centuries fell into ruin. Though in a few areas the very terraces constructed by the Incas are still in use today, this basic soil conservation technique has almost completely disappeared in the Andes.[23]

If the population of the Andes had remained at the reduced level of the seventeenth century, the absence of conservation practices would not be so threatening. With land abundant, farmers could afford to exploit the slope for a year or two, watch its topsoil and fertility wash away, and move on to another hill. But over the last century the population in the Andes has far surpassed that of the Inca Empire, with devastating consequences for the land and those whose livelihood depends on it.

Peru's total population was about four million at the turn of the century, nine million by 1950, and fifteen million by 1975. Colombia's population grew from about three million in 1900 to eleven million in 1950, and to twenty-six million in 1975. The ecological and social consequences of the mounting human pressure on the Andes become more apparent with each passing decade. Farmers are driven onto slopes so steep that erosion is a serious problem from the moment cultivation begins. Lands which need to rest for eight years, twelve years, or longer to regain their fertility can now be left fallow for only a few years.

A reduced fallow period results in the production of less organic matter and hence reduces the capacity of the soil to absorb and hold water. The soil structure deteriorates and compacts, resulting in an increase in the runoff of rainwater and an acceleration of erosion. An agricultural system that was viable under the conditions of earlier centuries is breaking down. It is no longer a rational adjustment to environmental conditions. Agricultural output is generally stagnant in the Andes, and in some areas is declining.[24]

Land reforms in Bolivia since 1953, and in Peru since 1964, are benefiting many mountain residents by improving the ownership patterns in these two countries but cannot, of course, create new land. Nor does land reform guarantee land-management practices that preserve the productivity of the soil, though it may be an essential first step. In the near-feudal economic system that has characterized the Andes, land redistribution is a prerequisite of improved farm productivity and modernized farming practices. But a look at Bolivia's mounting soil erosion problem in the decade and a half *after* the 1953 reforms suggests the necessity of safeguarding their benefits for future generations by the simultaneous introduction of new farming systems and family planning.

Overgrazing and overcropping have been depleting the fertility and vegetation of the Bolivian *altiplano* and the surrounding valleys at least since the beginning of Spanish rule, but fresh gulleys, increasingly frequent abandonment of once fertile fields, and a fall in crop yields on the steeper fields in recent years all indicate that the scale of damage has accelerated since the early fifties. David A. Preston argues that the increased access—for grazing, cropping, and firewood gathering—to areas once controlled by large landowners is an important cause of this acceleration in Bolivia, since this new access was not accompanied by a comprehension of soil conservation needs and practices. He writes:

> Their revolution effected a change in land ownership—the workers now own more land than the rich *hacendados* and they are free to make their own decisions, and they have their own representative organization. But nowhere has the social and political enfranchisement of the rural labour population brought a realization of the responsibilities involved with ownership, for such responsibilities have been ignored for 350 years. The Revolution has not changed the landscape: the peasants, freed by it, have made changes some of which, as we have demonstrated, have adversely affected the value of the land that they have so recently acquired.[25]

The mountainous third of Peru which, despite its harsh terrain, is the home of half the country's people, is entering into what R. F. Watters calls an "agricultural crisis." Deterioration may be most acute in the southern half of the *sierra*, as the mountain zone is called there. "This bleak, denuded region," Watters observes, "includes nearly one-half of the cultivated lands of Peru but produces only a little over one-sixth of the national income derived from agriculture."[26]

Families in the more densely settled areas of Peru and Bolivia, such as the Lake Titicaca Basin, have as little as one-half to two hectares of land available for farming. With the technologies currently employed, this area often does not provide enough food to

meet even the modest needs of the family. Chronic food shortages impel overexploitation of the soil and hence extensive soil
erosion, and finally result in the abandonment of farms and seasonal or permanent migration. Accentuating the erosion problem
in the most crowded areas of the *sierra* is a practice recently discussed in a Peruvian government study. Wood is so scarce that
not only is all available dung burned for fuel, but people pull the
roots out of the ground to burn as they cut the remaining
shrubs.[27]

In Colombia, erosion, landslides, and sedimentation—all natural
problems now accelerated by the pressures of humans and livestock
on the land—are major obstacles to national development. Cornell
University engineers working in one area, the Cauca region of the
southern mountains, have described the readily visible economic
damage, and the heavy loss of life, resulting each year from the
deteriorating environment. One recent landslide dammed the
Yumbo River and caused flooding deaths in the town of Yumbo.
"Landslides are so common that socially important slides occur every
few months," they write.

Recent deforestation in the watersheds has increased the intensity of floods and boosted the already heavy silt load of the region's
rivers. Much of this sediment is deposited where the mountain slopes
sink into valleys or plains. Cities, villages, and agricultural lands in
valleys are then damaged as the rivers spread or change their courses
due to the rising bed. Sediment deposits regularly block the Cali and
Canaveralejo rivers and cause major flooding problems in the city of
Cali. Rising silt loads may cost Colombia billions of dollars in hydroelectric benefits in coming decades. The lower Anchicaya Reservoir
has filled with silt in only seven years. Since the reservoir has lost its
capacity to store water, the multi-million-dollar hydroelectric plant it
was built to support now runs on the river flow alone—at one-third
of its planned capacity. A dredging operation keeps silt out of the
turbines.[28]

Colombia is not the only Andean country where a deteriorating
mountain environment is bringing destruction to the towns and
agriculture below. Increased flooding, the silting of riverbeds, and the
deposition of coarse, useless debris on fertile fields are common
phenomena throughout the Andean countries. In Venezuela, for

example, thousands of tons of detritus are deposited each year on fertile lands south of Lake Maracaibo by the rivers Catatumbo, Esculante, Chama, Motatán, and Carache.[29]

Deforestation upsets the normal workings of the hydrological cycle by affecting the course of fallen rains; instead of sinking into hillsides for later seepage into streams, rains on denuded lands run down in one major flood. Thus wet-season floods can increase while the flow in other seasons dries up entirely. Even in the nineteenth century, it was observed in Peru that several of the coastal rivers then flowing only seasonally had formerly carried a continuous flow. Denudation around their mountain sources had apparently caused the change. According to the Peruvian government study quoted above, deforestation in upper river valleys of the western slopes continues to disrupt river flows, and hence the valuable coastal irrigated agriculture of Peru.[30]

With life untenable for so many residents of the Andes, large-scale migration to other areas is inevitable.[31] As usually occurs when human populations are forced to flee an impossible situation, the movement out of the mountains has given rise to many new social problems. Some migration has been seasonal and keyed to available jobs, with able-bodied family members finding work in plantations, mines, or construction jobs along the coast or in Argentina. But the mushrooming shantytowns of the major cities of the Andean region attest to the fact that many uprooted families face conditions in their new homes little better than those they left. Unemployment, poverty, and despair are widespread.

The eastward movement of settlers to the jungles of the upper Amazon Basin is anarchic. Every government has colonization programs which have met with varying degrees of success or failure, but for the most part, the evacuees of the highlands are ill-prepared to face the unfamiliar conditions of the humid tropics. The frequent result is double disaster: a costly depletion of lumber resources as trees are cleared without regard to the soil's suitability for agriculture, followed shortly thereafter by the abandonment of the unproductive farms. Due to destructive farming techniques, notes R. F. Watters, "the migrants are re-creating many of the deplorable conditions from which they fled."[32]

There is, of course, no doubt that further resettlement of mountain residents in the lowlands will be necessary over the coming decades, as will the expansion of non-agricultural employment throughout the Andean region. The questions to be confronted are: How many people can the jungles and cities safely absorb? And will the spread of roads, infrastructure, and technical know-how in the eastern lowlands catch up with a continuing flow of settlers? Already, suggests Peru's Office for the Evaluation of Natural Resources, the most fertile agricultural soils of the country's vast jungle region are being cultivated. New colonists will have to settle on increasingly inferior soils.[33]

With the exception of the more economically developed Chile, the Andean countries have some of the world's highest population growth rates. At the current pace, the populations of Peru, Bolivia, Colombia, Ecuador, and Venezuela all will double in size inside of three decades. Oil, copper, and tin will certainly continue to boost the economic development and diversification of the Andean countries. Yet at this point it is difficult to imagine where these unborn citizens will live, where they will work, and where their food will grow.

Africa's principal mountainous zone, the highlands of the eastern side, presents a topography quite different from that of the Andes. These highlands, which stretch southward from Ethiopia through Kenya, Uganda, Tanzania, and beyond, are not a sharply rising mountain range in the customary sense, but are rather a wide, bulky massif. The huge Amhara Plateau, which constitutes most of Ethiopia and is the most extensive mountain region of any African country, rises abruptly from the surrounding arid plains and generally towers about two thousand meters above them. Yet the term "plateau" is also misleading in this case, for on these highlands are superimposed other mountains that rise to more than 4,500 meters above sea level. The plateau is ribboned by steep, sometimes impassable gorges and valleys, and is dotted with strange flat-topped mountains with nearly sheer sides—impregnable fortresses that have played a key role in the country's turbulent history. Jagged, intricate and fantastic, the Ethiopian highlands share with the Himalayas and Andes the unfortunate distinction of being one of the most erosion-prone areas on earth.

Ethiopia is remembered by many outsiders as the land whose late emperor at once stirred worldwide guilt and inspiration in

1936, when the League of Nations acquiesced in its invasion by Italy. Another special but frequently unappreciated role played by Ethiopia in the history of human civilization stems from its geography. It is commonly known that the cultural achievements of the ancient Egyptians were contingent upon the annual deposition of fertile silt on the River Nile's flood plain, but few have ever stopped to wonder where this silt came from. In nature few things are free, and for the bounty of the lower Nile, nature exacted a heavy price in the uplands of Ethiopia. Herodotus called Egypt the "gift of the Nile" but he might just as well have called it the gift of Ethiopia. The mud carried out of Ethiopia by the legendary Blue Nile and two other major tributaries of the Nile is a huge annual transfer of productive capacity.

With a population estimated at twenty-eight million in 1975, Ethiopia is the third most populous country in Africa, surpassed in numbers only by Nigeria and Egypt. The country's population growth rate of 2.6 percent annually is not so high as that of many neighboring countries. This is due not to a lower birth rate but to one of the world's highest death rates. Low agricultural productivity, and the historic need to give a high share of the land's produce to the landholding elite, has chronically kept much of the peasantry at a bare subsistence level.[34] In the bad years, such as those that visited the northern provinces of Wollo and Ogaden in the early and mid-seventies, the line between bare subsistence and famine for tens or even hundreds of thousands is quickly crossed.

The Ethiopian plateau receives good rainfall, and as much as three-fourths of the country was once forested. Clearing for cultivation, burning to create pasturelands, and tree felling to meet fuel and timber needs have reduced the forest area to a mere shadow of its historical domain; significant stands of timber now cover less than 4 percent of the country. The tempo of destruction has quickened since mid-century, and by the early sixties, reported an American research team, natural woodlands were disappearing at a rate of one thousand square kilometers per year. The country's major watersheds and steep mountain slopes are not being spared. "Even to the casual visitor to Ethiopia," a top-level soil conservation advisor recently wrote, "the extent of soil erosion seen in many parts of the country . . . will leave a lasting impression of desolation and impending disaster."[35]

The extent of deforestation and erosion varies by region according to population density and historical length of settlement. Leslie Brown, a prominent ecologist with decades of experience in East Africa, points out that one can observe the historical progress and results of land degradation in Ethiopia by journeying from north to south.[36] In the oldest inhabited areas to the north, such as the provinces of Tigre and Eritrea, some of the steepest slopes no longer even carry grass or shrubs. People coax what produce they can from eroded, infertile patches. Many streams have dried up except during the rainy season, when they are prone to violent torrents.

Following the depletion of the soils and forests to the north, the center of Ethiopian civilization moved southward around the tenth century.[37] The central highlands, including the region surrounding Addis Ababa, the nation's capital, have thus been increasingly exploited over the last millenium. Throughout much of this region, the high forest cover has been replaced by grass and scrub; springs have often dried up and silty rivers flow erratically. Farther to the south, where cultivators have more recently penetrated the forests, the streams tend to carry clear water and run constantly even in dry seasons.

A dramatic alteration in environmental quality has been visible within a single lifetime in the hills surrounding Addis Ababa. When the capital was founded in 1883 by the Emperor Menelik II, it was still surrounded by remnants of rich cedar forests and reasonably clear streams. Deforestation and erosion were immediately spurred by the influx of humans. In the ensuing nine decades, virtually all the available land in the region has been cultivated, while charcoal producers cut trees within a 160-kilometer radius for sale in the city. Now the waters of the nearby Awash River and its tributaries are thick with mud, and waterways are shifting their courses more markedly and frequently than in the past. A United Nations research team has expressed fears that the upper Awash Basin may become a "rocky desert."[38]

Within two decades after its founding, Addis Ababa faced a wood shortage so severe that some observers predicted the capital would have to be abandoned. And indeed it might have been, had not Menelik initiated a nationwide program to plant the eucalyptus tree, a quick-growing native of Australia that has spread to every continent. The eucalyptus saved Addis Ababa and is now the prime source of fuel and timber in many Ethiopian communities. Indeed, a single tree species has probably never had such a deep influence on the life of a city as the eucalyptus does in Addis Ababa today. Over 90 percent of its buildings have frames of eucalyptus wood. Wood and leaves from this species, gathered from trees inside the city, in plantations on its outskirts, and on farms in the surrounding countryside, are the city's principal source of fuel. A steady stream of porters, pack animals, and trucks bring eucalyptus wood and leaves into Addis Ababa daily.[39]

Unfortunately, the eucalyptus is not widespread enough in Ethiopia to eliminate the common practice of burning cow dung for fuel. The tree also has its disadvantages; it tends to dry out the soil wherever its numbers are profuse. Nor have eucalyptus plantations and copses proved a safeguard against erosion. The fallen bark and leaves of these trees do not readily decompose and mix in the soil, and the tendency for Ethiopians to systematically rake up available bark, twigs, and leaves under the trees for use as fuel further cuts their anti-erosion potential.

South of Addis Ababa in the Gamu highlands, Oxford University geographers recently witnessed the incipient breakdown of a sustainable agricultural system. When they visited this area in 1968, erosion was not yet a serious problem. The steeper slopes had been saved for grazing animals rather than plows, and the people showed an awareness of the erosion hazard by terracing the hillsides and constructing drainage channels on slopes to carry off excess rain. Animal manure was carefully collected and applied to the fields, while crop rotations and fallowing also helped to preserve the soil's fertility.

Under the pressure of a mounting population, however, farmers have started plowing up lands formerly reserved for grazing. This has accentuated overgrazing and, subsequently, erosion on the remaining pastures, and has also dented the cattle population. Fewer cattle mean less manure, which in turn means lower yields and greater requirements for arable land, which, to turn the screw again, will then necessitate further inroads into the pastureland, thus completing the cycle of degeneration. The villages in this area are violating their own land management rules and they know it, but they see no alternative.[40]

Ethiopia is unique in Africa for its seven and a half centuries of

virtually uninterrupted independence, and for its sixteen centuries of
Christianity. At least until the announcement of sweeping land reforms in early 1975 by the country's new military government, it also
had the most rigidly stratified land-tenure system on the continent.
Most of Ethiopia's best farmlands were owned by the Church, the
royal family, or the powerful landed aristocracy. The land was worked
by peasants in a state of bondage, whose daily lives were circum-
scribed by an elaborate system of social and economic castes.[41] There
is no question that the economic and political conservatism inherent
in this land-tenure system slowed the modernization of agriculture in
Ethiopia and reduced the incentive for the peasantry to manage soils
properly. Whether the new regime will successfully carry out its
announced reforms—and whether they will be accompanied by a
new concern for soil conservation—remain to be seen.

The fertile soils and pleasant climate of the highlands of Kenya,
Tanzania, and Uganda have proved attractive to both Africans and
Europeans, and it is no accident that Nairobi, Kenya's capital in the
heart of the highlands, is the nexus of industry and economic development in East Africa. Above undulating plains, deep valleys, and
numerous smaller mountain ranges tower Mount Kenya and Mount
Kilimanjaro, Africa's highest point.

These highlands are less rugged than those of Ethiopia, and
include some of the most productive—and densely settled—farmlands of Africa. They are bounded by zones infested with the tsetse
fly, a formidable enemy of the cattle so highly prized by most of East
Africa's peoples, and to the north by semi-arid lands best suited to
grazing. The concentration of people on the more fertile lands, such
as around Lake Victoria, on the slopes of Uganda's Mount Elgon, and
in the vicinity of Nairobi, is more than five hundred per square
kilometer—and in places it is far higher. In the colonial era, the
pressure of the African population on available lands in Kenya was
intensified by the reservation of a fourth of the arable area, including
many of the most fertile portions, for use by whites only. Since 1960,
these lands have been gradually resettled by Africans.

Between the efforts of land-hungry cultivators and charcoal mak-
ers, the East African highlands have been largely deforested except
for the most inaccessible mountain areas and occasional government-
protected reserves. Particularly where cultivators have moved up

steeper mountain slopes, and where the combination of population density and traditional techniques has run down fertility, erosion is on the rise.[42] Not surprisingly, migrants from the fragmented, over-crowded farming areas are pouring into the cities of East Africa, where they frequently wind up subsisting on whatever occasional work can be found. The future of East Africa's magnificent game reserves is also being jeopardized as the public pressure for new farmlands grows.

Ecological trends in the Uluguru Mountains of eastern Tanzania have been reasonably well-documented and provide a useful, if by now familiar-sounding, case study. The source of several important streams, the Ulugurus were once heavily forested, but, over the last century and a half, subsistence cultivation has spread throughout their slopes at the expense of trees. By the mid-twentieth century, the frightening consequences of denudation and improper cultivation practices in these mountains began to materialize. Severe flash flood-ing and silting have become expensive problems in downstream towns and croplands, and hydrological studies show flooding to be worsening. The River Ngerenge has begun to dry up completely in the dry season with increasing frequency. In 1960, for the first time, then again in 1966, sisal production on estates reliant on river waters suffered forced halts. By the late sixties, reports Paul H. Temple, a new and ominous phenomenon was recorded in the Ulugurus—catastrophic landslide damage, a trend which elsewhere has been characterized as the terminal phase of human-induced accelerated erosion.[43]

All in all, little hard data is yet available on the scale and nature of environmental deterioration in the mountains. However, there is a rather broad consensus among the various scientists and govern-mental agencies cited above, as well as many others, about the general direction of prevailing mountain trends. On the basis of already available knowledge, it is no exaggeration to suggest that many moun-tain regions could pass a point of no return within the next two or three decades. They could become locked in a downward spiral from which there is no escape, a chain of ecological reactions that will permanently reduce their capacity to support human life. A hastened flow of migrants to urban slums will be a principal consequence.

This possibility is very real, but it is not inevitable. There is no major mountain problem for which technological solutions are not already known. If the existing negative trends are not abruptly reversed within the coming decades, it will be because human institutions have failed to adapt themselves to environmental necessity.

While every mountain zone has its peculiar problems and solutions, some general considerations with wide application can be noted. Undoubtedly, the most important need in virtually every case is an intensification of food production on the best farmland—in the lowlands and valleys, and on the gentler mountain slopes. Only when this occurs can the self-defeating pressure to move onto ever steeper hillsides be countered. In many cases reforms in land tenure and the distribution of extension and credit services—political, not technical tasks—are the first requirements of agricultural progress.

Where hillsides must be farmed, the adoption of soil-conservation techniques is essential. Often erosion can be curbed markedly through the simplest of measures. In parts of Nepal, for example, many terraces are poorly constructed, with an outward rather than inward slope and an inadequate buttress of stone to help them survive the annual monsoon deluge. In the Andes, a restoration of the lost art of terracing would bring immediate benefits to farmers, the land, and cities downstream. Terracing is not always the answer; soil experts in the Uluguru Mountains of Tanzania, for example, found that terracing encourages landslides there and exposes too much infertile subsoil. They did find that strategic tree planting, farming on the contour, and other simple changes in cultivation practices could greatly reduce erosion.[44] Measures like these have not been widely adopted because farmers are either not aware that they are possible or are not convinced that sufficient production benefits would ensue to justify the extra labor. Many mountain farmers have never seen an agricultural extension agent.

Where population pressures do not permit a return of mountain slopes to forest, which might be the ecological ideal, the introduction of permanent tree crops like apples, apricots, nuts, or timber plantations may be a good compromise. Tree crops combine many of the ecological advantages of the forest with employment and income for former farmers; an apparently successful United Nations watershed improvement project in Pakistan has utilized foreign-donated food aid as wages in road construction and planting activities to help tide farmers over until their new orchards start producing income.[45]

More extensive reforestation programs are needed throughout the mountains of Asia, Latin America, and Africa. Trees are required not only to protect vulnerable slopes and soils, but also to provide firewood and thus reduce the increasing use of manures for cooking fuel. Putting manure back onto the fields will in turn help boost their productivity, reducing the pressure to spread cultivation onto unsuitable slopes. Virtually every government in the mountain regions has demarcated forest reserves in especially strategic locations such as those above important rivers. But it is only when adequate food and fuel are available from other sources that these "protective forests" can be genuinely protected.

Greater opportunities to earn a living outside of agriculture can also reduce pressures on the land. Mining and related industries are already a source of jobs and money in the Andes. However, the environmental consequences of these operations must be carefully monitored and controlled, lest their impact prove self-defeating. It has become something of an axiom in many quarters that the poor countries cannot afford the luxury of pollution controls on their industries, but the gap between environmental protection needs in the rich and poor countries may be narrower than many think. In Peru, a government agency points out that air pollution is killing vegetation on thousands of mountainous hectares surrounding mines and refineries, resulting in "truly spectacular" soil erosion.[46]

Tourism, too, at once poses a great potential and a threat for the mountains. With their fascinating scenery and cultures, countries like Nepal, Peru, and Ethiopia clearly can expand their tourist trade severalfold. Yet planning will be essential to prevent the further degradation of their natural resources by visiting sightseers. The soaring number of trekkers in the high Himalayas of Nepal over the last decade has created a booming firewood business for some mountain people, but it has grown at the expense of the forests and particularly fragile ecosystems of the upper slopes.

The central threat to the future of the mountains is the burden of the burgeoning human numbers they must bear. Planned migration to less crowded lowlands, where possible, will be important, but can only purchase a brief reprieve. The need to rapidly bring population growth to a halt in the mountains cannot be circumvented; the limited carrying capacity of these lands will become all too plain over the next few decades.

It is generally easy to recommend technological answers to ecological problems. Political and cultural factors are invariably the real bottlenecks holding up progress. Changing the relationship of people to land in the mountains, as anywhere else, invariably involves sensitive changes in the relationship of people to one another. Developmental funds and talents spent in the mountains are resources denied the cities and the plains. In the end, the greatest challenge of all may be convincing the people of the plains that the future of the mountains cannot be isolated from their own.

6. The Other Energy Crisis: Firewood

Dwindling reserves of petroleum and artful tampering with its distribution are the stuff of which headlines are made. Yet for more than a third of the world's people, the real energy crisis is a daily scramble to find the wood they need to cook dinner. Their search for wood, once a simple chore and now, as forests recede, a day's labor in some places, has been strangely neglected by diplomats, economists, and the media. But the firewood crisis will be making news—one way or another—for the rest of the century.

While chemists devise ever more sophisticated uses for wood, including cellophane and rayon, at least half of all the timber cut in the world still serves in its original role for humans—as fuel for cooking and, in colder mountain regions, home heating. Nine-tenths of the people in most poor countries today depend on firewood as their chief source of fuel. And all too often, the growth in human population is outpacing the growth of new trees—not surprising when the average user burns as much as a ton of firewood a year.[1] The results are soaring wood prices, a growing drain on incomes and physical energies in order to satisfy basic fuel needs, a costly diversion of animal manures from food production uses to cooking, and an ecologically disastrous spread of treeless landscapes.

The firewood crisis is probably most acute today in the countries of the densely populated Indian subcontinent and in the semi-arid stretches of central Africa fringing the Sahara Desert, though it plagues many other regions as well. In Latin America, for example, the scarcity of wood and charcoal is a problem throughout most of the Andean region, Central America, and the Caribbean.

As firewood prices rise, so does the economic burden on the urban poor. One typical morning on the outskirts of Kathmandu, Nepal's capital city, I watched a steady flow of people—men and women, children and the very old—trudge into the city with heavy, neatly chopped and stacked loads of wood on their backs. I asked my taxi driver how much their loads, for which they had walked several hours into the surrounding hills, would sell for. "Oh wood, a very expensive item!" he exclaimed without hesitation. Wood prices are a primary topic of conversation in Kathmandu these days. "That load costs twenty rupees now. Two years ago it sold for six or seven rupees." This 300 percent rise in the price of fuel wood has in part been prompted by the escalating cost of imported kerosene, the principal alternative energy source for the poor. But firewood prices have risen much *faster* than kerosene prices, a rise that reflects the growing difficulty with which wood is being procured. It now costs as much to run a Kathmandu household on wood as on kerosene.

The costs of firewood and charcoal are climbing throughout most of Asia, Africa, and Latin America. Those who can, pay the price, and thus must forego consumption of other essential goods. Wood is simply accepted as one of the major expenses of living. In Niamey, Niger, deep in the drought-plagued Sahel in West Africa, the average manual laborer's family now spends nearly one-fourth of its income on firewood. In Ouagadougou, Upper Volta, the portion is 20 to 30 percent.[2] Those who can't pay so much may send their children, or hike out into the surrounding countryside themselves, to forage—if enough trees are within a reasonable walking distance. Otherwise, they may scrounge about the town for twigs, garbage, or anything burnable. In many regions, firewood scarcity places a special burden on women, who are generally saddled with the tasks of hiking or rummaging for fuel.

In some Pakistani towns now, people strip bark off the trees that line the streets; thus, meeting today's undeniable needs impoverishes the future. When I visited the chief conservator of forests in Pakistan's North West Frontier Province at his headquarters in the town of Peshawar, he spoke in a somewhat resigned tone of stopping his car the previous day to prevent a woman from pulling bark off a tree.

"I told her that peeling the bark off a tree is just like peeling the skin off a man," he said. Of course the woman stopped, intimidated by what may be the most personal encounter with a senior civil servant she will have in her lifetime, but she doubtless resumed her practice shortly, for what else, as the chief conservator himself asked, was she to do?

It is not in the cities but in rural villages that most people in the affected countries live, and where most firewood is burned. The rural, landless poor in parts of India and Pakistan are now feeling a new squeeze on their meager incomes. Until now they have generally been able to gather free wood from the trees scattered through farmlands, but as wood prices in the towns rise, landlords naturally see an advantage in carting available timber into the nearest town to sell rather than allowing the nearby laborers to glean it for nothing. This commercialization of firewood raises the hope that entrepreneurs will see an advantage in planting trees to develop a sustainable, labor-intensive business, but so far a depletion of woodlands has been the more common result. And the rural poor, with little or no cash to spare, are in deep trouble in either case.

With the farmland trees and the scrubby woodlands of unfarmed areas being depleted by these pressures, both the needy and the ever-present entrepreneurs are forced to poach for fuel wood in the legally protected, ecologically and economically essential national forest reserves. The gravity of the poaching problem in India has been reflected in the formation of special mobile guard-squads and mobile courts to try captured offenders, but law enforcement measures have little effect in such an untenable situation. Acute firewood scarcity has undermined administrative control even in China, where trees on commune plantations are sometimes surreptitiously uprooted for fuel almost as soon as they are planted.[3]

Trees are becoming scarce in the most unlikely places. In some of the most remote villages in the world, deep in the once heavily forested Himalayan foothills of Nepal, journeying out to gather firewood and fodder is now an *entire day's* task. Just one generation ago the same expedition required no more than an hour or two.[4]

Because those directly suffering its consequences are mostly illiterate, and because wood shortages lack the photogenic visibility of famine, the firewood crisis has not provoked much world attention.

And, in a way, there is little point in calling this a world problem, for fuel-wood scarcity, unlike oil scarcity, is always localized in its apparent dimensions. Economics seldom permit fuel wood to be carried or trucked more than a few hundred miles from where it grows, let alone the many thousands of miles traversed by the modern barrel of oil. To say that firewood is scarce in Mali or Nepal is of no immediate consequence to the boy scout building a campfire in Pennsylvania, whereas his parents have already learned that decisions in Saudi Arabia can keep the family car in the garage.

Unfortunately, however, the consequences of firewood scarcity are seldom limited to the economic burden placed on the poor of a particular locality, harsh as that is in itself. The accelerating degradation of woodlands throughout Africa, Asia, and Latin America, caused in part by fuel gathering, lies at the heart of the profound challenges to environmental stability and land productivity reviewed in this book—accelerated soil erosion, increasingly severe flooding, creeping deserts, and declining soil fertility.

The Dust Bowl years in the Great Plains of the thirties taught Americans the perils of devegetating a region prone to droughts. The images provided by John Steinbeck in *The Grapes of Wrath* of the human dislocation wrought by that interaction of people, land, and climate could easily describe present-day events in large semi-arid stretches of Africa along the northern and southern edges of the Sahara, and around the huge Rajasthan Desert in northwest India. Overgrazing by oversized herds of cattle, goats, and sheep is the chief culprit, but gathering of fuel wood is also an important contributor to the destruction of trees in these regions. Firewood is a scarce and expensive item throughout the Sub-Saharan fringe of Africa, all the way from Senegal to Ethiopia, but citizens in towns like Niamey are paying a much higher price than they realize for their cooking fuel. The caravans that bring in this precious resource are contributing to the creation of desert-like conditions in a wide band below the desert's edge. Virtually all trees within seventy kilometers of Ouagadougou have been consumed as fuel by the city's inhabitants, and the circle of land "strip-mined" for firewood—without reclamation—is continually expanding.

In the Indian subcontinent, the most pernicious result of firewood scarcity is probably not the destruction of tree cover itself, but the alternative to which a good share of the people in India, Pakistan, and Bangladesh have been forced. A visitor to almost any village in the subcontinent is greeted by omnipresent pyramids of hand-molded dung patties drying in the sun. In many areas these dung cakes have been the only source of fuel for generations, but now, by necessity, their use is spreading further. Between three hundred and four hundred million tons of wet dung—which shrink to sixty to eighty million tons when dried—are annually burned for fuel in India alone, robbing farmland of badly needed nutrients and organic matter. The plant nutrients wasted annually in this fashion in India equal more than a third of the country's chemical fertilizer use. Looking only at this direct economic cost, it is easy to see why the country's National Commission on Agriculture recently declared that "the use of cow dung as a source of non-commercial fuel is virtually a crime." Dung is also burned for fuel in parts of the Sahelian zone in Africa, Ethiopia, Iraq, and in the nearly treeless Andean valleys and slopes of Bolivia and Peru, where the dung of llamas has been the chief fuel in some areas since the days of the Incas.[5]

Even more important than the loss of agricultural nutrients is the damage done to soil structure and quality through the failure to return manures to the fields. Organic materials—humus and soil organisms which live in it—play an essential role in preserving the soil structure and fertility needed for productive farming. Organic matter helps hold the soil in place when rain falls and wind blows, and reduces the wasteful, polluting runoff of chemical nutrients where they are applied. Humus helps the soil absorb and store water, thus mitigating somewhat the impact on crops of drought periods. These considerations apply especially to the soils in tropical regions where most dung is now burned, because tropical topsoils are usually thin and, once exposed to the burning sun and torrential monsoon rains, are exceptionally prone to erosion and to loss of their structure and fertility.

Peasants in the uplands of South Korea have adopted a different, but also destructive way to cope with the timber shortage. A United Nations forestry team visiting the country in the late 1960s found not only that live tree branches, shrubs, seedlings, and grasses were cut for fuel; worse, many hillsides were raked clean of all leaves, litter, and burnable materials. Raking in this fashion, to meet needs for home fuel and farm compost, robs the soil of both a protective cover and organic matter, and the practice was cited by the UN experts as "one of the principal causes of soil erosion in Korea." Firewood scarcity similarly impairs productivity in eastern Nigeria, where the Tiv people have been forced to uproot crop residues after the harvest for use as fuel. Traditionally, the dead stalks and leaves have been left to enrich the soil and hold down erosion.[6]

The increasing time required to gather firewood in many mountain villages of Nepal is leading to what the kingdom's agricultural officials fear most of all. For once procuring wood takes too long to be worth the trouble, some farmers start to use cow dung, which was formerly applied with great care to the fields, as cooking fuel. As this departure from tradition spreads, the fertility of the hills, already declining due to soil erosion, will fall sharply. In the more inaccessible spots, there is no economic possibility whatsoever of replacing the manure with chemical fertilizers.

And so the circle starts to close in Nepal, a circle long completed in parts of India. As wood scarcity forces farmers to burn more dung for fuel, and to apply less to their fields, falling food output will necessitate the clearing of ever larger, ever steeper tracts of forest— intensifying the erosion and landslide hazards in the hills, and the siltation and flooding problems downstream in India and Bangladesh.

Firewood scarcity, then, is intimately linked in two ways to the food problem facing many countries. Deforestation and the diversion of manures to use as fuel are sabotaging the land's ability to produce food. Meanwhile, as an Indian official put it, "Even if we somehow grow enough food for our people in the year 2000, how in the world will they cook it?"

B. B. Vohra, a senior Indian agricultural official who has pushed his government ahead on numerous ecological causes, shook his head as we talked in his New Delhi office. "I'm afraid that we are approaching the point of no return with our resource base. If we can't soon build some dramatic momentum in our reforestation and soil conservation programs, we'll find ourselves in a downward spiral with an irresistible momentum of its own." Without a rapid reversal of prevailing trends, in fact, India will find itself with a billion people to support and a countryside that is little more than a moonscape. But the politicians, in India and other poor countries, will take notice when they realize that, if people can't find any firewood, they will surely find something else to burn.

The firewood crisis is in some ways more, and in others less, intractable than the energy crisis of the industrialized world. Resource scarcity can usually be attacked from either end, through the contraction of demand or the expansion of supply. The world contraction in demand for oil in 1974 and early 1975, for example, helped to ease temporarily the conditions of shortage.

But the firewood needs of the developing countries cannot be massively reduced in this fashion. The energy system of the truly poor contains no easily trimmable fat such as that represented by four- to five-thousand-pound private automobiles. Furthermore, a global recession does little to dampen the demand for firewood as it temporarily does in the case of oil. The regrettable truth is that the amount of wood burned in a particular country is almost completely determined by the number of people who need to use it. In the absence of suitable alternative energy sources, future firewood needs in these countries will be determined largely by population growth. Firewood scarcity will undoubtedly influence the urgency with which governments address the population problem in the years ahead.

Even if the demographers are surprised by quick progress on the population front over the next few decades, the demand for basic resources like firewood will still push many countries to their limits. Fortunately trees, unlike oil, are a renewable resource when properly managed. The logical immediate response to the firewood shortage, one that will have many incidental ecological benefits, is to plant more trees in plantations, on farms, along roads, in shelter belts, and on unused land throughout the rural areas of the poor countries. For many regions, fast-growing tree varieties are available that can be culled for firewood inside of a decade.

The concept is simple, but its implementation is not. Governments in nearly all the wood-short countries have had tree-planting programs for some time—for several decades in some cases. National forestry departments in particular have often been aware of the need to boost the supply of wood products and the need to preserve forests for an habitable environment. But several problems have generally plagued tree-planting programs.

One is the sheer magnitude of the need for wood and the scale of the growth in demand. Population growth, which surprised many with its acceleration in the post-war era, has swallowed the moderate tree-planting efforts of many countries, rendering their impact almost negligible. Wood-producing programs will have to be undertaken on a far greater scale than most governments presently conceive if a significant dent is to be made in the problem.

The problem of scale is closely linked to a second major obstacle to meeting this crisis: the perennial question of political priorities and decision-making time-frames. With elections to win, wars to fight, dams to build, and hungry people to feed, it is hard for any politician to concentrate funds and attention on a problem so multi-dimensional and seemingly long-term in nature. Some ecologists in the poor countries have been warning their governments for decades about the dangers of deforestation and fuel shortages, but tree-planting programs don't win elections. This phenomenon is, of course, quite familiar to all countries, not just the poorest. In the United States, resource specialists pointed out the coming energy crisis throughout the 1960s, but it took a smash in the face in 1973 to wake up the government, and, as of late 1975, the country still can hardly be said to have tackled the energy challenge head-on.

Despite these inherent political problems, India's foresters made a major breakthrough a few years back as the government drew up its five-year development plan for the mid-to-late seventies. Plans were laid for the large-scale establishment of fast-growing tree plantations, and for planting trees on farms and village properties throughout the country.[7] A program is going ahead now, but there have been some unexpected events since the projects were first contemplated two or three years ago: the quintupling of the world price of petroleum, the tripling in price and world shortages of grains and fertilizers, and the subsequent wholesale diversion of development funds to maintenance and emergencies in order to merely muddle through 1974 without a major famine and a total economic breakdown. India's development efforts have been set back several years by recent events, and forestry programs have not been immune to this trend.

Even with the right kind of political will and the necessary allocation of funds, implementing a large-scale reforestation campaign is a complex and difficult process. Planting millions of trees and success-fully nurturing them to maturity is not a technical, clearly defined task like building a dam or a chemical-fertilizer plant. Tree-planting projects almost always become deeply enmeshed in the political, cultural, and administrative tangles of a rural locality; they touch upon, and are influenced by, the daily living habits of many people, and they frequently end in failure.

Most of the regions with too few trees also have too many cattle, sheep, and goats. Where rangelands are badly overgrazed, the leaves of a young sapling tempt the appetites of foraging animals. Even if he keeps careful control of his own livestock, a herdsman may reason that if his animals don't eat the leaves, someone else's will. Maraud-ing livestock are prime destroyers of tree-planting projects through-out the less developed world. Even if a village is internally disciplined enough to defend new trees from its own residents, passing nomads or other wanderers may do them in. To be successful, then, reforesta-tion efforts often require a formidable administrative effort to protect the plants for years—not to mention the monitoring of timber-harvesting and replanting activities once the trees reach maturity.

Village politics can undermine a program as well. An incident from Ethiopia a few years back presents an extreme case, but its lessons are plain. A rural reforestation program was initiated as a public works scheme to help control erosion and supply local wood needs. The planting jobs were given to the local poor, mostly landless laborers who badly needed the low wages they could earn in the planting program. Seedlings were distributed, planting commenced, and all seemed to be going well—until the overseers journeyed out to check the progress. They found that in many areas the seedlings had been planted upside down! The laborers, of course, well knew the difference between roots and branches; they also knew that given the feudal land-tenure system in which they were living, most of the benefits of the planting would flow one way or another into the hands of their lords. They were not anxious to work efficiently for substand-ard wages on a project that brought them few identifiable personal returns.[8]

In country after country, the same lesson has been learned: tree-planting programs are most successful when a majority of the local community is deeply involved in planning and implementation, and clearly perceives its self-interest in success. Central or state governments can provide stimulus, technical advice, and financial assistance, but unless community members clearly understand why lands to which they have traditionally had free access for grazing and wood gathering are being demarcated into a plantation, they are apt to view the project with suspicion or even hostility. With wider community participation, on the other hand, the control of grazing patterns can be built into the program from the beginning, and a motivated community will protect its own project and provide labor at little or no cost.

An approach like this—working through village councils, with locally mobilized labor doing the planting and protection work—is now being tried in India. There are pitfalls; Indian villages are notoriously faction-ridden, and the ideal of the whole community working together for its own long-term benefit may be somewhat utopian. But if it can get underway on a large scale, the national program in India may succeed. Once given a chance, fast-growing trees bring visible benefits quickly, and they just could catch on. The Chinese have long used the decentralized, community labor-mobilizing approach to reforestation, apparently with considerable success.

A new program of global public education on the many benefits of reforestation, planned by the United Nations Environment Program, will hopefully direct more attention to the tremendous global need for tree planting. Whatever the success of reforestation projects, however, the wider substitution of other energy sources for wood can also contribute greatly to a solution of the firewood problem. A shift from wood-burning stoves to those running on natural gas, coal, or electricity has indeed been the dominant global trend in the last century and a half. As recently as 1850, wood met 91 percent of the fuel needs of the United States, but today in the economically advanced countries, scarcely any but the intentionally rustic, and scattered poor in the mountains, chop wood by necessity anymore. In the poor countries, too, the proportion of wood users is falling gradually, especially in the cities, which are usually partly electrified, and where residents with any income at all may cook their food with bottled gas or kerosene. Someone extrapolating trends of the first seven decades of this century might well have expected the continued spread of kerosene and natural gas use at a fairly brisk pace in the cities and into rural areas, eventually rendering firewood nearly obsolete.

Events of the last two years, of course, have abruptly altered energy-use trends and prospects everywhere. The most widely over-looked impact of the fivefold increase in oil prices, an impact drowned out by the economic distress caused for oil-importing coun-tries, is the fact that what had been the most feasible substitute for firewood, kerosene, has now been pulled even farther out of reach of the world's poor than it already was. The hopes of foresters and ecologists for a rapid reduction of pressures on receding woodlands through a stepped-up shift to kerosene withered overnight in Decem-ber, 1973, when OPEC announced its new oil prices. In fact, the dwindling of world petroleum reserves and the depletion of wood-lands reinforce each other; climbing firewood prices encourage more people to use petroleum-based products for fuel, while soaring oil prices make this shift less feasible, adding to the pressure on forests.

The interconnections of firewood scarcity, ecological stress, and the broader global energy picture set the stage for some interesting, if somewhat academic, questions about a sensible disposition of world resources. In one sense it is true that the poor countries, and the world as a whole, have been positively influenced by the OPEC countries, which, through price hikes and supply restrictions, are forcing conservation of a valuable and rapidly disappearing resource, and are not letting the poor countries get dangerously hooked on an undependable energy supply. In a world in which energy, ecology, and food were sensibly managed, however, the oil distribution picture would look far different. The long-term interest in preserving the productive capacity of the earth and in maximizing welfare for the greatest number of people might argue for lower prices and a rapid *increase*, not a halt, in the adoption of kerosene and natural gas in the homes of the poor over the next two decades. This, in turn, would be viable for a reasonable time period only if the waste and compara-tively frivolous uses of energy in the industrial countries, which are depleting petroleum reserves so quickly, were cut sharply. It is not so far-fetched as it might first seem to say that today's driving habits in Los Angeles, and today's price and production-level decisions in the Persian Gulf, can influence how many tons of food are lost to floods in India, and how many acres of land the Sahara engulfs, in 1980.

Fossil fuels are not the only alternate energy source under consid-eration, and, over the coming decades, many of those using firewood, like everyone else, will have to turn in other directions. Energy sources that are renewable, decentralized, and low in cost must be developed. Nothing, for example, would be better than a dirt-cheap device for cooking dinner in the evening with solar energy collected earlier in the day. But actually developing such a stove and introducing it to hundreds of millions of the world's most tradition-bound and penniless families is another story. While some solar cookers are already available, the cost of a family unit, at about thirty-five to fifty dollars, is prohibitive for many, since, in the absence of suitable credit arrangements, the entire amount must be available at once. Mass production could pull down the price, but the problem of inexpensively storing heat for cloudy days and evenings has not been solved.[9] The day may come in some countries when changes in cooking and eating habits to allow maximum use of solar cookers will be forced upon populations by the absence of alternatives.

Indian scientists have pioneered for decades with an ideal-sounding device that breaks down manures and other organic wastes into methane gas for cooking and a rich compost for the farm. Over eight thousand of these bio-gas plants, as they are called, are now being used in India. Without a substantial reduction in cost, however, they will only slowly infiltrate the hundreds of thousands of rural villages, where the fuel problem is growing. Additionally, as the plants are adopted, those too poor to own cattle could be left worse off than ever, denied traditional access to dung, but unable to afford bio-gas.[10] Still, relatively simple and decentralized devices like solar cookers and bio-gas plants will likely provide the fuel sources of the future in the poor countries.

In terms of energy, Nepal is luckier than many countries in one respect. The steep slopes and surging rivers that cause so many environmental problems also make Nepal one of the few remaining countries with a large untapped hydroelectric potential. The latent power is huge, equaling the hydroelectric capacity of Canada, the United States, and Mexico combined. Exploitation of this resource will be expensive and slow, but will relieve some of the pressures being placed on forests by the larger towns of Nepal and northern India. On the other hand, cheap electricity will only partly reduce firewood demands, since the electrification of isolated villages in the rugged Himalayas may never be economically feasible.

The firewood crisis, like many other resource problems, is forcing governments and analysts back to the basics of human beings' relationship with the land—back to concerns lost sight of in an age of macro-economic models and technological optimism. Awareness is spreading that the simple energy needs of the world's poorest third are unlikely ever to be met by nuclear power plants, any more than their minimum food needs will be met by huge synthetic-protein factories.

Firewood scarcity and its attendant ecological hazards have brought the attitude of people toward trees into sharp focus. In his essay "Buddhist Economics," E. F. Schumacher praises the practical as well as esoteric wisdom in the Buddha's teaching that his followers should plant and nurse a tree every few years.[11] Unfortunately, this ethical heritage has been largely lost, even in the predominantly Buddhist societies of Southeast Asia. In fact, most societies today lack a spirit of environmental cooperation—not a spirit of conservation for its own sake, but one needed to guarantee human survival amid ecological systems heading toward collapse.

This will have to change, and fast. The inexorable growth in the demand for firewood calls for tree-planting efforts on a scale more massive than most bureaucrats have ever even contemplated, much less planned for. The suicidal deforestation of Africa, Asia, and Latin America must somehow be slowed and reversed. Deteriorating ecological systems have a logic of their own; the damage often builds quietly and unseen for many years, until one day the system falls asunder with lethal vengeance. Ask anyone who lived in Oklahoma in 1934, or Chad in 1975.

7. The Salting and Silting of Irrigation Systems

THE FIRST human civilization may well have emerged in what is now Iraq, on the plain of the Tigris and Euphrates rivers known as Mesopotamia. Here people learned, six or seven thousand years ago, to raise abundant crops in the desert by diverting river water onto their fields. The need to maintain irrigation works, in turn, provided strong impetus for cooperation among families and clans, while the new food surplus encouraged formation of the first non-farming classes—priests, administrators, warriors, merchants, craftsmen, and artists.

That Mesopotamia is possibly the world's oldest irrigated area is not an encouraging observation. The end result of six millenia of human management is no garden spot. The region's fertility was once legendary throughout the Old World, and the American conservationist Walter Lowdermilk has estimated that at its zenith Mesopotamia supported between seventeen and twenty-five million inhabitants.[1] One early visitor, Herodotus, wrote that "to those who have never been in the Babylonian country, what has been said regarding its production will be incredible." Today Iraq has a population of ten million, and on the portions of these same lands that have not been abandoned peasants eke out some of the world's lowest crop yields.

Rarely has the history of civilization been so directly linked to changing ecological conditions. Environmental modification has sometimes been the cause, and sometimes been the result, of the rise and fall of Mesopotamian states. Their history reveals lessons for

irrigation-based societies everywhere, lessons only slowly being absorbed despite the visual aid of Iraq's wastelands.

The first recorded civilization, that of the Sumerians, was thriving
in the southern Tigris-Euphrates valley by the fourth millenium B.C.
Over the course of two thousand years Sumerian irrigation practices
ruined the soil so completely that it has not yet recovered. The
dissolved salts present in minute quantities in all water do not evaporate, so as water evaporates or is used by plants, the salt content of
the remaining water grows. Irrigation waters unused by plants and
not absorbed by the air may percolate down to the underground water
table, which, in turn, may become increasingly salty over time. In
Mesopotamia, in the absence of natural or artificial means of draining
away extra water, the water table began to rise. Seepage from canals
and periodic floods also added to the groundwater until it eventually
reached root zones, where its brackishness stunted or prohibited crop
growth. Even where the subterranean water was not too salty, it
damaged or killed crop roots by cutting off the oxygen supply.

Once the water table was within several feet of the surface, water
also began to move upward by capillary action, pulled by the dry air
and soil above. As it reached the surface and evaporated, a thin
deposit of salts was left behind—a familiar enough phenomenon to
anyone who has seen a glass or dish from which salty water has
evaporated. Centuries of repeated wetting and drying out left a thick
white crust on the ground, rendering much of it all but useless for
agriculture. Vast areas of southern Iraq today glisten like fields of
freshly fallen snow; from 20 to 30 percent of the country's potentially
irrigated land is unusable. Where salt concentrations are lower, farmers in much of Iraq are able to farm a given area only in alternate
years, despite the availability of ample water. Irrigation water is
applied to wash the salts downward into the ground a modest distance, a small crop of salt-tolerant barley is coaxed from the earth,
and then the land is given over to deep-rooted weeds for a year, which
help dry out the soggy soils. Until an underground system of drainage
tiles is installed—an expensive, ambitious proposition—the situation
will not improve; it may deteriorate further.

The progressive build-up of salt in lower Meospotamia has been
carefully chronicled by archaeologists. Barley tolerates saltier water
and soils than wheat, so as salinization proceeded, barley has grown

over larger areas. About 3500 B.C., wheat and barley were grown in nearly equal proportions. A thousand years later, wheat accounted for only one-sixth of the crop, and eight hundred years after that, by 1700 B.C., wheat cultivation had ceased entirely in most of southern Iraq.

More important than the shift in crops was the decline in productivity, a process which accelerated as irrigation grew heavier in the waning years of the Sumerian civilization. The declining yields were devastating for the towns, where large non-farming classes were dependent on farmers' surpluses for sustenance. Field records from 2400 B.C. indicate an average yield of 2,537 liters, or about two metric tons, of grain per hectare—respectable by current North American standards. By 2100 B.C. the yield had declined to 1,460 liters per hectare, and by 1700 B.C., as political and cultural dominance shifted northward to Babylonia, yields in Sumer had fallen to 897 liters per hectare. Many of the great Sumerian cities dwindled to villages or were left in ruins. While "probably there is no historical event of this magnitude for which a single explanation is adequate," write the pre-eminent historians Thorkild Jacobsen and Robert M. Adams, "that growing soil salinity played an important part in the breakup of Sumerian civilization seems beyond question."[2]

Successive Mesopotamian civilizations grew up in the north, where salinity generally did not emerge as a problem until many centuries later, and has never been as acute as in the lower reaches of the rivers. But as irrigation systems grew in size and complexity, another threat to food production eventually grew as well: the burden of silt carried by the river waters into irrigation canals. For the first four thousand or so years, irrigation in Mesopotamia was a simple affair in comparison to its later phases; silt was not yet a serious problem. Rather small human settlements concentrated along the banks of major water courses meant that short, easily maintained canals were the principal engineering works needed to supply fields with water. Often, river bank levees could simply be breached to flood low-lying fields.

After Alexander the Great's conquest in the fourth century B.C., Mesopotamia entered a new phase of urbanization and agricultural development. These trends peaked in the third through seventh centuries A.D., when the conquering Sassanians of Persia built an unprecedented, gigantic system of major and minor canals that crisscrossed the desert, allowing the exploitation of almost all the region's cultivable area. One ambitious canal, the Nahrwan, was three hundred kilometers in length with thousands of brick sluice gates along its branches.

The Tigris and Euphrates both carry heavy loads of sediment from upstream, loads that quite likely increased over time due to the growing use and misuse of the rivers' headwater areas for farming and grazing. The deposition of silt on floodplains, riverbeds, and at the rivers' delta on the Persian Gulf is a massive and never-ending process. Thus, over the last five thousand years a new layer of soil ten meters thick has been laid over the landscape in some parts of the Tigris-Euphrates Basin. In the earlier, simpler phase of irrigated agriculture and settlement in Mesopotamia, silting must have been an occasional nuisance when it clogged channels and ditches, but, on the whole, it was a blessing since, in the pattern of Egypt's productive floodplain agriculture, it meant an annual deposit of fertile topsoil on the land.

The new scale and complexity of agriculture in the Sassanian period, however, also meant a new vulnerability to any factor that might disrupt the workings of the extensive, finely-tuned water delivery system. The long canals quickly filled with sediment, and keeping the system working required continuous and backbreaking drudgery. Coordination of labor over a large area was required; no village acting in isolation could ensure its own survival. A powerful, centralized state was essential. That the system worked for many centuries is attested to by the huge mounds of silt, once cleaned from canals by slaves and laborers, that are still a central feature of the landscape in parts of central and northern Iraq.

The region flourished so long as state power remained inviolate, and the Sassanian cities became cosmopolitan trade centers where Hebrew, Greek, and Syriatic were commonly spoken. But whenever cooperative canal maintenance was not forthcoming, whether due to internal social decay, a preoccupation with external military adventures, or hostile invasion, the system suffered catastrophic breakdowns that reduced the region's life-support capacity. A crisis of political authority in the late Sassanian period caused just such a breakdown, and, after a brief recovery following the Islamic invasion, the irrigation system fell into a state of disrepair in the twelfth century that has not yet been righted. Accumulating silt loads reduced the flow of the Nahrwan Canal to a trickle, and a leading world civilization to a few dusty villages. Most historians now agree that the Mongols, who invaded the region in the thirteenth century and have often since been blamed for its downfall, found scenes of devastation rather than opulence when they arrived in the legendary land of Mesopotamia.

A central lesson of Iraq's history, one that has been learned and forgotten repeatedly since the decline of the Sumerians, is that irrigation (providing water to thirsty crops) and drainage (removing excess water from the soil) are inseparable components of a single system. The need is not simply to locate water and wet the earth, though this goal alone has, understandably, preoccupied all desert societies, but to provide moisture at a level that maximizes crop growth and preserves the land's productive capacity over time. Often, this means providing underground or surface drains to draw away water unused by plants. Ironically, once water, the prerequisite of production, is obtained, getting rid of it eventually becomes nearly as crucial as assuring its continued supply.

Some irrigated regions are blessed with adequate natural underground drainage or an efficient flow of surface water out of fields. In Egypt, for example, the annual flooding of the Nile flushed salts out of the soil each year, which is why the Nile Valley, in contrast to Mesopotamia, has remained one of the world's most productive and densely populated areas for several thousand years. Egyptian irrigation projects outside the Nile floodplain have, over the last century, developed severe salinity problems. The Aswan Dam, which harnessed the Nile in the sixties, will allow further extensions of the irrigated area, but has eliminated the historic natural soil desalination process of the Nile Valley. Waterlogging and salinity, on both old and new farmlands, are becoming major headaches for the Egyptian government, and coping with them demands tremendous expenditures and technical skill.

Either too much or too little irrigation water can cause salinity problems. Where underground drainage is poor, as in Iraq, unused and unevaporated irrigation waters build up toward the surface, causing the dual problems of waterlogging and, as the water near the ground evaporates, salt deposits. But when water applications are too scanty, salt also builds up, since not enough water filters downward to flush the inevitable salt deposits below the root zone.

Under the right conditions, typified by Pakistan, the problems of too much and too little water can, in a sense, both exist at once. As the human population of Pakistan has multiplied over the last century, the irrigated area has also grown—but without a commensurate increase in the amount of water extracted from the country's rivers and underground aquifers. As a result, water is spread thinly and is often inadequate to satisfy the needs of crops, let alone the need to flush or leach salt downward out of the surface soils. At the same time, however, natural underground drainage throughout much of Pakistan is poor, and seepage from the country's extensive network of unlined canals, together with irrigation waters unused by crops, have brought the water table almost to the surface in many areas.

Pakistan, whose Indus Plain is by far the largest continuously irrigated region on earth and is thus a major world resource, presents many parallels to ancient Mesopotamia. It is a desert land where all civilization is founded on the waters of the Indus and its four major tributaries. Salt and silt challenge the survival of Pakistani society just as they did those of the Sumerians and Sassanians. One advantage the Pakistanis hold over the ancients is access to modern technologies that can probably keep irrigation sustainable on a permanent basis. A disadvantage is a population of seventy million that promises to double in twenty-five years, leaving little latitude for mistakes or disruptions in the water supply without tragic consequences.

Irrigation in Pakistan has a long history, stretching back to the Indus civilization of over four thousand years ago. Water diversions on a large scale, however, are relatively recent. Canal irrigation began at the end of the seventeenth century, primarily to furnish water for the parks and gardens of the Moghul royalty. Though the obvious agricultural potential of these canals, most of which could only operate when river flows were high, was quickly perceived, farming was by and large still limited to narrow ribbons of land near the riverbanks. In the mid-nineteenth century the British undertook construction of the massive perennial irrigation system that was to make the area called the Punjab the breadbasket of the Indian subcontinent. Low dams called barrages were built to store water and channel it into canals with a total length of tens of thousands of miles, some of which carry nearly as much water as the Colorado River of the southwestern United States.

Almost as soon as these canals were built something unexpected began to happen: the level of water in wells began to rise. The new water courses were altering the equilibrium of the hydrological cycle, with as much as a third of the water they carried seeping down to the subterranean water table. Away from the riverbeds the water table was initially fifty to seventy feet deep, but over huge areas the table began to rise steadily, at a rate of one or two feet per year, until it was within ten or fifteen feet of the surface. At this point its rise slowed as the capillary rise and evaporation of water helped reduce the pace of build-up. But in some areas the water table eventually reached the surface, emerging as stagnant puddles. Well before the water reached the roots of the crops, damage to the land's productivity began occurring. Once underground water with a salinity content of one part per thousand—quite acceptable to crops—is within several feet of the surface, capillary evaporation residues will raise the salt content of the top three feet of soil to the intolerable level of one part per hundred in just twenty years.

By 1960 the problems of waterlogging and salinity were taking a deadly toll. An estimated area of over two million hectares, a fifth of the annually cultivated area of the Indus Plain, was severely affected; either yields were significantly cut by waterlogging and/or salinity, or production had ceased altogether. As many as forty-thousand additional hectares were falling into that category each year, a good share of them lost to cultivation altogether. And the productivity of many more millions of hectares was well below its potential level due to saline soils.[3] Pakistan was losing a hectare of good agricultural land every twenty minutes, but gaining a new claimant on that land by birth every twenty-four seconds. Pakistan's existence as a nation was being called into question by this process of human-caused ecological change, and the question could no longer be tabled.

Until 1954, almost nothing was done to counter the threat posed by waterlogging and salinity. Research efforts initiated in that year culminated in a comprehensive strategy for attacking the problem that was launched by the government in 1961. Warning that Pakistan had to act soon to avoid a "national catastrophe," and invoking the spectre of ruined hydraulic civilizations of the past, the government called for the construction of tens of thousands of miles of drains to draw surplus water away from the fields. But the heart of the new program, particularly in the Punjab, was the massive installation of tubewells, cylindrical shafts driven down to tap the water table.[4] Water drawn from underground could, engineers reasoned, replace or supplement river waters in irrigating the fields. Part of the water drawn up by the wells would seep back down, flushing salts downward, and enough would be lost to evaporation or plant consumption so that the dangerously high subterranean water would gradually sink lower. Where conditions were right, the introduction of properly spaced tubewells could not only help reverse the waterlogging and salinity trends, but could also provide a large, dependable supply of new irrigation water, allowing the adoption of higher-yield grain varieties.

Salinity-control project areas were given boundaries and thousands of wells were sunk. The results were significant, if sometimes below the expectations of the program designers. Tests in late 1964 in the initial project area indicated that in one year, despite unusually heavy rainfall, the groundwater level had fallen by eleven inches.[5] Overall, by 1970, it appeared to many observers that an important threshold had been crossed, with the land reclamation by then outstripping land losses.

But the processes that produce waterlogging and salinity continue without reprieve, while competing claims on the attention and resources of the government have left unrealized the salinity program's original goals for tubewells and drain construction. While the cost of the recently built Tarbela Dam on the Indus River climbed from the original estimate of $400 million to $1.2 billion, the civil war that resulted in the breaking away of Bangladesh also soaked up funds needed to fight salinity. Thus it would be premature to claim that the salinity problem has been conquered by Pakistan. As one leading foreign advisor to the government recently said, "The battle is far from won, but at least it is now well underway. The country appears to be holding ground." In 1973 President Bhutto publicly recognized the shortcomings of the past effort. "So far we have merely toyed with the problems," he said, as he announced a new plan involving the anticipated expenditure of half a billion dollars over the next decade to hold down the losses to salinity.[6]

Irrigation in Pakistan, like that in ancient Iraq, is also hampered by the extraordinarily high silt load carried by the Indus and its tributaries—the product of easily erosive geologic conditions, exacerbated by the progressive deforestation of the Himalayas. The principal threat posed to Pakistan by silt is not the asphyxiation of irrigation canals, bad as that problem may be, but the shockingly rapid sedimentation of the country's expensive new reservoirs. Pakistan's topography offers dam sites that are poor by usual international standards, and, where feasible at all, dams will produce reservoirs that are extremely small in comparison with the size of the rivers they harness. The combination of small reservoirs and large rivers bearing heavy silt loads means reservoirs will have very brief life-spans.

The country's first big dam project, the $600 million Mangla Reservoir, began operations in 1967. It was constructed on the basis of feasibility studies showing a reservoir life expectancy of one hundred years or more. Sediment measurements after a few years of operation, however, indicate that most of its water-holding capacity will be gone in fifty-five to seventy-five years.[7] Even this estimate makes no allowance for a possible acceleration in silting as deforestation and overgrazing in the Jhelum River watershed continue.

The engineering lengths to which water-short, densely populated countries can be forced is suggested by Pakistan's Tarbela Dam, which was completed in early 1975. The largest earth- and rock-filled dam in the world, Tarbela was constructed with certain knowledge that its reservoir will be almost completely filled with silt in fifty years.[8] The reservoir's extraordinary expense is nevertheless outweighed in the calculations of the Pakistani government by its benefits in electric power and irrigation. As the original reservoir loses its water storage capacity, water will be diverted to nearby valleys, and eventually another dam may be built downstream. In the case of the Mangla Reservoir, plans now stand eventually to heighten the dam, thus adding to its storage ability and postponing the demise of the irrigation system it supports.

One generation of costly concrete and steel structures can be replaced by another, but as yet little has been done to attack accelerating soil erosion in the Indian and Pakistani headwaters of the Indus and its tributaries. Pakistan is paying an enormous price for land deterioration in the western Himalayas. Any reduction in the flow of soil into the Indus system would bring life-giving benefits to the country by prolonging the economic life of these reservoirs.

Continuously threatened by waterlogging, salinity, and siltation, Pakistan presents a striking juxtaposition of the primitive and the ultra-modern. Many of the world's poorest people subsist oblivious to advanced technologies, and their livelihood and habits are little affected by the decisions of seemingly faraway national governments. In Pakistan, however, the survival of an extremely poor, largely uneducated population is increasingly dependent on sophisticated technologies, elaborate regional cooperation, and huge, well-planned governmental expenditures. Fighting the salinity problem requires complex analysis using computers, and the careful control of pumping and irrigation practices over large areas by highly trained technicians. The country's huge irrigation projects demand centralized control and cooperation over wide areas; in the face of a system breakdown the individual farmer would be just as helpless as the Mesopotamian farmers were who lost their livelihood nine hundred years ago. Effectively replacing the new reservoirs as they fill with silt will require tremendous expenditures and technical skill. Should prolonged political disintegration or military conflict ever impair for long the government's ability to meet these problems forcefully, Pakistan's resemblance to Mesopotamia would leave little to the imagination.

Only a few decades back, it was sometimes debated whether the irrigation of desert lands, where evaporation is particularly high and natural drainage is often poor, could be considered a permanent enterprise. Today, most experts believe the technology is available to maintain soil quality indefinitely. Yet, right up to the present day, world irrigation development has been blemished by the frequent neglect of this knowledge in the planning and construction of new irrigation projects. Where drainage works have been provided at all, they have often been inadequate.

Insufficient provision for drainage in many countries is one aspect of a larger problem plaguing world irrigation development. Opening the initial stages of a project such as a dam or major canal is always a prestigious event, with high political payoffs for those responsible. These are visible, impressive achievements constructed and thereafter managed by a small core of technological elites. But the extra steps required to make the grand projects meaningful and sustainable—the development of systems to deliver water on time and in proper quantities to the individual farmer, and the assurance of proper drainage at the farm level—are often overlooked. These steps are more difficult, requiring the cooperation not only of a few highly skilled technicians, but of all.

One explanation for this self-defeating trend is the mindset of many civil engineers, who sometimes plan and build dams with too little appreciation for the management needs of the individual farmer and for the long-term requisites of soil quality. Irrigation is too often viewed as an end in itself, rather than as the handmaiden of agricultural development. There may, furthermore, be an anti-drainage bias built into the process of project evaluation and adoption. Government officials and agencies competing for scarce funds and status, not to mention the consultants and contractors who will design and build the dams and canals, have a strong stake in keeping overall cost estimates as low as possible, and in magnifying a project's potential benefits. It is common at the planning stage to underestimate both the need for and the cost of the other side of the irrigation coin, drainage. Once irrigation projects are under construction and costs run over initial estimates, as they usually do, there are additional temptations to postpone even the planned drainage work. This is particularly easy since salinity and waterlogging usually do not start undercutting production for many years after the irrigation, with its dramatic initial returns, gets underway—and well after construction firms and bureaucrats have moved on to other projects, countries or positions.[9]

The legacy of this continued defiance of reality is a stupendous loss of global agricultural output. Two American analysts have estimated that the productivity of at least one-third of all the world's irrigated land is being undermined to varying degrees by salinity problems. The noted Soviet soil scientist V. Kovda goes further, estimating that 60 to 80 percent of all irrigated lands are, due to inadequate drainage or canal lining, becoming gradually more saline and, hence, infertile. By his calculations, twenty to twenty-five million hectares of land have been laid waste over the centuries after the introduction of improperly managed irrigation, and two hundred thousand to three hundred thousand additional hectares—out of a total worldwide irrigated area of nearly two hundred million hectares —pass from cultivation *each year* due to waterlogging and salinity.[10] Strangely, the awesome dimensions of this worldwide problem are generally unrecognized.

While Pakistan and Iraq present the most dramatic examples of the salinization threat, the problem is a critical one on every inhabited continent. Waterlogging and salinization are reducing yields to varying degrees in virtually every one of the thirty or so countries with more than half a million hectares of land under irrigation.

In Pakistan's neighbor, India, for example, over six million hectares—compared to a total national irrigated area of about forty million hectares—have been severely damaged by waterlogging and salinity and in many cases rendered unusable. Governmental and private tubewell installation has recently helped draw down the water table in parts of the Indian Punjab, but there and elsewhere, in the states of Haryana, Uttar Pradesh, Gujarat, Maharashtra, Orissa, and Rajasthan, the threat remains acute. After learning bitter lessons from past irrigation projects, the Indians are carefully minding drainage needs as the waters of the huge new Rajasthan Canal bring life to the desert, though some analysts still fear that salinity will harm the newly irrigated lands. China, too, has suffered from improper irrigation practices. Yields on at least a fifth of the irrigated area of some major regions are reduced by salinity, and tens of millions of hectares throughout the country are barren because of salt—some of it natural, some of it left by farmers in the past.[11]

Waterlogging and salinity plague nearly every country in the Middle East, not just Iraq. Salt-whitened sands are surpassed only by oil derricks as noteworthy human alterations of the landscape in that part of the world. As irrigation development escalates in the arid countries of the Middle East and North Africa, in part due to the new capital resources being shared with neighboring states by the oil exporters, salinity damage is sure to increase.

In the Euphrates Valley of Syria, above the Euphrates' entry into Iraq, from one-fourth to one-half the total agricultural area has been rendered unfit for cultivation by soil salinity and saturation. Average cotton yields in the valley's remaining farmlands dropped from 2.5 tons per hectare in the early fifties to about 1.5 tons per hectare by 1966. More than half of the combined irrigable lands of the Euphrates and Khabour Valleys, about 220,000 hectares, had been harmed or destroyed by salinity as of 1970. Nearly all this damage has occurred in the last twenty-five years since the introduction of intensive, perennial cotton production.[12] The problem is certain to spread with the current expansion of irrigated area allowed by Syria's new dam on the Euphrates.

In 1970, less than a decade after Jordan initiated irrigated agriculture in the Jordan River Valley, salt and sogginess were affecting 12 percent of the project area, and the extent of the injury, according to the government, was "increasing every year." In Iran, officials say that "the majority of irrigated lands are saline and crop yields are depressed by the toxcity of salt." Iran is planning to bring vast quantities of water that is already moderately saline to large areas of land; only conscientious attention to drainage and leaching needs will permit these investments to fulfill their potential.[13]

Waterlogging and salinity problems also affect every country with major irrigation works in the Americas. By 1200 A.D., as the Incas of the mountains above were gaining ascendency over Peru's coastal cultures, some irrigated lands of the coastal desert had already been abandoned—due, some historians feel, to salinity and a rising water table. Today over two hundred thousand hectares along Peru's coast, almost a third of the cultivated area in this region, suffer from the same problems. Twenty to thirty percent of the agricultural lands in the Patagonian region of Argentina have been damaged by salt accumulation. In Northeast Brazil, at least half the irrigated land is affected by waterlogging and salinity, and the yields on some irrigated fields are actually lower now than they were before the advent of irrigation. Food and Agriculture Organization researchers have also identified salinity as an acute and growing problem in northwestern Haiti, a region suffering from drought and famine in mid-1975. As salt accumulates on the fields there, the cultivation of important food crops like peanuts and beans is impaired and then abandoned. More salt-resistant crops like cotton are substituted until their yields, too, begin falling; eventually the land is given over to cactus and thorny brush. Leaching and drainage could reverse this catastrophic trend, report the FAO analysts, and the battle is now getting underway.[14]

As a result of inadequate provision for drainage, and the attempt to spread irrigation waters too thinly, the tremendous expansion in this century of irrigation systems in Northwestern Mexico has been plagued by serious salinity problems.[15] In the Yaqui Valley, for example, forty thousand hectares were damaged by salts and fifteen thousand hectares had been retired from production by the mid-sixties. The problem is even worse in the large irrigated zone of the Colorado Delta, where four-fifths of the arable land was to some degree affected, and 14 percent was too saline for any cultivation by 1965. The Mexican government has recently initiated drainage projects in many salt-damaged areas, but another basic problem—the effort to farm too much land that is often alkaline with too little water— ensures that the salinity threat will not be countered easily.

Exacerbating the already difficult salinity problem in Mexico's Colorado Delta, the salt content of the lower Colorado River leaving the United States nearly doubled after 1961. In 1944 the precious waters of the Colorado were divided between the United States and Mexico, but no attention was given the question of water quality. The river has a high natural salt content due to geologic deposits, and, as its waters have been increasingly diverted for agricultural uses and then drained back into the river, the salt concentration downstream has steadily risen. Since irrigation uses part of the water and speeds evaporation, the salt content in leftover waters rises.

The salt content of the river escalated dramatically when a pumping program was initiated in 1961 to lower the highly saline water table in the Wellton-Mohawk Irrigation and Drainage District of Arizona. Brackish water caused serious agricultural damage downstream in Mexico's Mexicali Valley, and became a major diplomatic irritant between Mexico and the United States. Protracted negotiations finally produced an agreement in 1973, under which the United States is building a desalination plant in Yuma, Arizona, to improve the quality of the water drained from the Wellton-Mohawk District.[16]

Though the 1973 accords were billed as a "final solution" to this international dispute over water quality, the problem persists. Irrigation development in the southwestern United States continues, and the expected diversion of upper Colorado waters for oil shale and coal exploitation promises to intensify further the salinity problems downstream.[17] Since the Colorado is the life-blood of much of the southwestern United States, there is little choice but to use and re-use the river waters, even though each additional diversion adds to the salt concentration of the remaining flow.

Already, despite elaborate drainage systems, some farmers in the rich Imperial and Coachella Valleys of California have suffered falling crop yields. The drainage facilities are not yet as developed in the Mexicali Valley, but the future of irrigated agriculture on both sides of the border will be thorny in any case as the salt content of the river water rises, a process beyond the control of any single farmer or region. In the lower Colorado Basin, writes Harold E. Dregne, a prominent U.S. agronomist, "it is not hard to see the day coming when on-the-farm ameliorative measures are no longer sufficient."[18] Southern California, Arizona, and northwestern Mexico are gradually being boxed into a predicament with no easy answers. Either the river's salt build-up will have to be reversed in the coming decades —perhaps in part through massive, expensive desalination operations, but also by cutting off some of the upstream salt sources—or large areas of cropland will eventually have to be abandoned. There are limits to how many unnatural burdens can be heaped upon a single river, a fact whose implications many individuals now recognize, but which the affected state and national governments have so far been unable to reflect in practice.

Silt, like salt, is undermining irrigation systems on every continent. Heavy sedimentation of canals and other waterworks is a common problem, reducing the efficiency of irrigation systems and absorbing funds and labor for maintenance that are sorely needed for activities to increase production. The accelerated silting of reservoirs, cutting their effectiveness for hydroelectric-power generation, flood control, and irrigation, has huge annual worldwide costs that have never been totaled.

Silt in rivers and streams is, of course, one of the major end results of erosion on the land. The natural conditions of vegetation and soil structure in the watershed area, and of riverbanks and bottoms, are usually the major determinants of sediment loads carried by waterways. But human alteration of the landscape can also make a significant difference, giving once-clear streams a visibly muddy complexion. Excess sediment is the major form of human-caused water pollution in the world today, and exacts a heavier cost than any other water pollutant—possibly more than all other pollutants combined.

Sediment studies of the U.S. Department of Agriculture in six-teen Mississippi watersheds carried out in 1959–61 examined the potential impact of various human land uses on silt loads.[19] The studies were in a hilly region where, once the original forest cover is stripped away, soils are especially prone to erosion. The annual soil runoff from forested areas was only a few hundredths of a ton per hectare. Erosion from grassy pasturelands was much higher, though still not dangerous, averaging four tons per hectare. Lands cultivated for corn production, however, produced an average annual sediment yield of fifty-four tons per hectare. On the more skillfully managed fields, the erosion rate was eight tons per hectare, but on some farms it reached 106 tons. Where gullies had formed on abandoned farm-lands, the average runoff of sediment was an astounding 450 tons per hectare. The numbers differ in every region but the basic relation-ships do not: as forests are cleared for agriculture, or are severely depleted by lumbering, grazing, or firewood gathering, the silt load of the rivers and streams below usually rises.

Siltation is especially dramatic in reservoirs that are small in relation to the river feeding them, such as the Mangla and Tarbela reservoirs in Pakistan. The water-holding capacity of the Tarbela Reservoir is only about one-seventh the annual flow of the soil-laden Indus, ensuring a short life. Lake Mead on the Colorado River, by contrast, holds more than twice as much water as the Colorado carries in a year, and will remain useful for centuries, despite the river's high sediment load.

The threat of siltation is frequently underestimated by dam build-ers, resulting in unhappy surprises later on for governments and farmers. When construction began on the Anchicaya Dam in Co-lombia in 1947, the consulting engineers reassured apprehensive offi-cials that "tropical rivers carry little sediment," though the influx of farmers and roads to the surrounding slopes was then beginning. Imagine the bewilderment of these officials in 1957 when, only twenty-one months after the dam was completed, nearly a fourth of the reservoir's capacity had been lost to sediment![20]

Preliminary calculations for Taiwan's Shihmen Reservoir sug-gested a useful life of seventy-one years, but in one five-year period, from 1963 to 1968, 45 percent of its capacity was lost, provoking a new government program to halt unauthorized forest clearing and the rapid spread of farming onto steep slopes upstream. The Philippines' Ambuklao Dam, built to last sixty-two years, will be useful for only thirty-two years—mainly because of excess logging in the Upper Agno River watershed.[21] The literature of world development activities is strewn with similar horror stories—and there will be more.

One of West Africa's most significant recent development projects, the Kainji Reservoir, which opened in 1969 on the Niger River in northern Nigeria, is intended to spur industrial and agricultural development in that arid zone by providing cheap electricity and water. Unfortunately, satellite photographs analyzed by Paul Anaejionu of the American University, in Washington, D.C., suggest that the project will not meet original expectations. Deforestation, burning and overly intensive cultivation in Mali, Niger, Upper Volta, Dahomey, and Nigeria are apparently dumping far more sediment into the river than anticipated.[22]

Surveying twenty-two irrigation reservoirs, the Indian government recently discovered, to its consternation, that in many instances the annual inflow of sediment is at least four times as high as was assumed when the dams were built. In some cases, laments B. B. Vohra of the Ministry of Agriculture, "there are no alternative sites for reservoirs once these have been rendered useless. This means that even if we have the money to build fresh projects—and this is by no means certain—we shall not have the physical opportunities for doing so."[23] A good reservoir site is, in a sense, a non-renewable resource. Once exploited and then destroyed by silt, it can be replaced only at increased expense—and sometimes, as in some unlucky Indian provinces, it is irreplaceable.

A sample survey in 1941 indicated that 39 percent of the reservoirs in the United States had a useful life of less than fifty years. "Many of the reservoirs," write two prominent engineers, "were doubtless designed without consideration of the sediment load normally carried by the streams on which they were built." The total economic loss from reservoir sedimentation in the United States was estimated at fifty million dollars per year as of 1962.[24] Watershed conservation programs are not cheap, but they will often pay for themselves many times over in the extra farm output allowed by prolonged reservoir life.

Irrigation has played a key role in the evolution of Southeast Asia, where wet-rice production has permitted the emergence of the world's most densely peopled humid tropical zones. Among Southeast Asian countries, irrigation agriculture is perhaps most advanced in Java, the island on which nearly two-thirds of the people of Indonesia live. Population growth is forcing other countries in that part of the world, such as the Philippines and Thailand, in Java's direction —toward more elaborate irrigation networks—while Java rehabilitates its long-neglected system in the effort to boost yields, and further extends the intricate circulatory network of canals that permits extra harvests each year on its scarce arable lands.

Many observers fear that Java is rushing inexorably toward an ecological nightmare. A few numbers sketch the problem. By 1975, Java, with a land area of 136,000 square kilometers, or about that of the state of New York, had a population of eighty-four million people, giving it an overall density of over six hundred per square kilometer (sixteen hundred per square mile). This compares with a density in the crowded Netherlands of about four hundred per square kilometer —and in some of Java's intensively farmed rural areas the density is more than double the island's average. Java's population is growing by two million per year, but there is no ecologically viable possibility of adding to the cultivated lands, which now cover 60 percent of the island.

Even if average fertility drops by one-fourth, reports Sumitro Djojohadikusumo, a leading Indonesian analyst, Indonesia's total population will increase from about 130 million in 1975 to 250 million at the century's end, and Java's from 84 million to 146 million. This will give Java an *average* density of about eleven hundred people per square kilometer. As Sumitro observes, "this is greater than the present density in the most populated and urbanized centres of Western Europe."[25] Java is rapidly becoming the world's largest "island city." The resettlement of Javanese on Indonesia's more sparsely populated islands is desirable and necessary, but can hardly alter Java's basic prospects. Even if the government's goal of moving two million people during the current Five-Year Plan is met, Java's annual population increase will only be reduced by a fifth. There are basic logistical limits as to how many people can be resettled rapidly, particularly since they are moving to territories that are, in many cases, agriculturally untested and ecologically unfamiliar. The steady flow of destitute squatters into the shantytowns of Jakarta remains the country's most notable form of migration.

The term "island city" brings to mind an image of a fairly prosperous commercial center like Hong Kong or Singapore. Java, however, is an agricultural land where the vast majority must support themselves through farming if they are to support themselves at all. Simply finding housing space for the island's residents without encroaching on good farmlands will be a major challenge. Incredibly, Indonesia is not yet a massive food importer; limited progress in raising rice yields, and the wide sharing of inadequate food supplies among the rural poor, mean chronic undernourishment but, as yet, no huge import needs. And one can identify the theoretical means by which Java could continue to limp along with a nutritionally inadequate, but probably politically tolerable, supply of food as its population doubles. The efficiency of the island's irrigation system can be increased greatly, and this, together with improved seeds and other practices, could bring major increases in crop yields. Fisheries development can help boost protein supplies, and the exploitation of as yet untouched arable lands on other islands of Indonesia may provide a surplus to help keep Java afloat.

But well before any theoretical agronomic limits are reached, Java could find itself sliding into an ecological crisis that undercuts their meaning. Farmers, out of understandable desperation, have *already* cleared forests well beyond the safe limits of agricultural sprawl, denuding hillsides on which farming is not sustainable for long. The negative consequences can be calculated in terms of disrupted stream flows and rising silt loads that jeopardize the efficiency of the island's essential water-delivery systems. Satellite pictures suggest that as little as 12 percent of this lush island now has tree cover, including plantations of teak and other species. In the watersheds of the vital Solo, Brantas, and Citarum river systems, forest cover is certainly well under 10 percent. An Indonesian ecologist, Dr. Otto Soemarwoto, has documented the acceleration of soil erosion in several watersheds, drawing special attention to a sevenfold increase in the silt load of the Citarum over a recent period of three years that is rapidly filling up Indonesia's largest reservoir downstream at Jatiluhur. In some areas around the upper Citarum, cassava and other root crops are now planted right up to the tops of smaller peaks that were covered by dense jungle in the 1950s.[26]

As Java's population grows, pressures on the ecosystem will mount, but as the irrigation system is refined, it is becoming more and more vulnerable to ecological disruption. Java's irrigation system might be compared to a fine watch. From some vantage points, one can see waters running in nine different directions and levels. Closely synchronized water movements from rivers or reservoirs to the individual field are essential; maximizing yields requires a precise amount of water at just the right time, and then its removal from fields.

Any factor reducing the efficiency of this intricately meshed system will pull down crop yields. A growing load of silt in canals and gates, like blood clots in human arteries, impairs their effectiveness. Declining river flows in dry seasons, and higher floods in the wet seasons, which are additional consequences of hillside clearing, further hamper the functioning of the system. An ominous long-term trend is the gradual building up of the level of fields and ditches as they receive a heavy annual sediment deposit, hindering the delivery and drainage of water to and from fields by gravity.

As other Southeast Asian countries expand and intensify their irrigation facilities to meet burgeoning food needs, moving in the direction of what might be termed the "Java model," their vulnerability to siltation and disrupted river flows will grow commensurately. The Philippines, for example, hopes to triple its irrigated area over the next decade. Yet the watersheds of several of the country's important rivers have already been ravaged by shifting cultivators and loggers. Satellite photographs indicate that deforestation is far more advanced than official statistics reveal, with the forest cover probably less than a fifth of the country's land area—a far cry from the 35 to 50 percent commonly assumed.

In mainland Southeast Asia, many of the same factors operate, but generally on a different schedule. The relatively crude irrigation system in the basin of Thailand's Chao Phya River has been vastly upgraded in the past decade, with large reservoirs on upstream tributaries and intricate water-delivery systems that permit two crops annually on much of the Central Plains area below. Hydrological data show little change in silt loads and seasonal water flows over the past decade, while aerial and satellite photos show more than 40 percent forest cover remaining in the Chao Phya watershed.

Yet some local observers calculate that the northern forests are being decimated at a rate of 5 to 7 percent a year. Most of the deforestation is occurring in the mountainous homelands of the minority ethnic tribes, especially in the "slash-and-burn" areas of opium cultivation—regions where the authority of the central Thai government is minimal. Superficially, the Thai situation appears relatively comfortable and well-balanced, like Java's in the 1930s. But if one projects a continued 3 percent annual rate of population growth, and a 5 to 7 percent annual deforestation rate, a crisis point is coming in the near future: intensive, controlled irrigation will be most needed just at the time when upstream erosion and irregularity of water flow will make downstream delivery and drainage facilities most difficult to operate and maintain. And this point will most likely coincide with greatly sharpened competition for Chao Phya water to serve agricultural, urban, industrial, and hydroelectric needs. A large river flow must also be maintained to keep ocean water from invading the low-lying coastal plains.

Though Java has a refined and intricate irrigation system already facing acute stresses, and Thailand a less pressured system with a grace period of perhaps fifteen years in which to undertake integrated planning, the basic lesson remains the same. The expected benefits of irrigation projects may never materialize, in Southeast Asia or elsewhere, if more heed is not paid to the overall ecological balance.

Each of the irrigation systems in ecological danger discussed in this chapter represents a serious local threat to a region or nation. Viewed together against world trends in irrigation development, they take on a far broader significance. One of the key factors permitting world food output to keep up with the surging postwar demand has been the historically unprecedented explosion in irrigation capacity. A total world irrigated area of eight million hectares in 1800 reached forty million hectares in 1900, 105 million hectares in 1950, and then 190 million hectares by 1970, thus growing faster than world population so far this century.

Unquestionably, considerable potential remains for further growth in the total world irrigated area, not to mention the vast improvements possible in the efficiency of water use. Large reserves of underground water remain untapped in the Indian subcontinent and elsewhere, and the Sudan has not come close to utilizing the Nile River waters accorded it by treaty with Egypt, to take two notable examples. Yet, on a global basis, there are clear indications that the rate of new irrigation development will slow markedly over the coming decades.[27] Most of the world's economically feasible opportunities for large-scale irrigation development have already been exploited. In some areas, such as the high plains of West Texas, the underground water supplies are being steadily depleted.

The great burst in world irrigation projects during the third quarter of this century cannot be repeated during the last quarter. Thus, though world irrigated area expanded by nearly 3 percent annually between 1950 and 1970, it will probably grow at little more than 1 percent a year in the remaining years of the century. As the need for fresh water grows faster than its availability, the insidious loss of irrigation capacity to salt and silt is sure to become more visible.

8. Myth and Reality in the Humid Tropics

THE TROPIC OF CANCER and the Tropic of Capricorn bound mostly open sea, but the oceans are interrupted by the bulk of Africa and Latin America and the southern tail of Asia. While half this tropical land mass is actually grassland, the landscape conjured in the minds of most people by the word "tropics" is that of the lushly vegetated humid zones—the dense forests of the Amazon Basin, Central and West Africa, and Southeast Asia. These evergreen forests now cover about a fourth of the tropics, and an unknown share of the grasslands would also be in forest had human beings not intervened with fire.

In parts of Southeast Asia such as Java, rich volcanic soils and centuries of careful construction support perennial rice cultivation and compact human settlements. In the humid tropical areas of every continent, there are alluvial plains (composed of topsoil washed down by rivers) that are potentially quite productive and capable of supporting dense populations. What is characteristic of most rainy regions in Latin America and Africa, however, is the mere handful of people they support.

The sparse population of much of the humid tropics is puzzling on its face, for, in terms of sheer mass, no zone produces more life. Tropical forests are the most diverse ecosystems on earth, with thousands of plant species—many still unrecorded by modern science—crowded into small areas. Viewing what appears to be the gross underutilization of the humid tropics by farmers who slash fields out of the forest only to abandon them and move on after a few years, some analysts have pinned great hopes on these zones as possible suppliers of food to a hungry world. With year-round sunlight, warmth, and rainfall, the huge potential of the humid tropics to produce vegetation is obvious. What still eludes the human race is the means to harness much of that potential for its own ends.

After a century strewn with dashed dreams and squandered investments, the myth of boundless fertility in the tropics is gradually dying. Remarks made by United Nations soil scientists in 1965 on the Latin American experience apply equally to large areas of tropical Africa and Asia:

> It is no accident that much of the yet unfarmed land in Latin America lies within the humid tropics, for here are to be found the soils where failure to understand the dynamic nature of the soil system brings swift disaster. The whole history of Man's penetration of these regions has been chequered by high hopes followed by failure, and those who have remained have often chosen to emulate the shifting cultivation system of the indigenous farmers. Few indeed are the examples of successful, efficient and permanent agricultural industry established in the humid tropics, and most of these are concerned with special crops on especially favored soils.[1]

In most temperate farmlands, mineral nutrients accumulate in the soil for potential use by plants. In the tropical forests, by contrast, most nutrients are locked in the profuse, multi-layered vegetation. The soil is less a nutrient source than a mechanical support for plant life which constitutes an almost closed cycle of growth and decay. When the land is stripped of its dense cover, the soil temperature soars under the intense equatorial sun, hastening biological activity and the deterioration of remaining organic matter. Torrential downpours, sometimes bringing six to eight inches of rain in a single day, wash away the thin topsoil and leach the scant nutrients it holds downward, beyond the range of crop roots.

Tropical forest residents have developed a special tactic known as "shifting cultivation" for coaxing sustenance out of their land.[2] The forest vegetation is cut away to form a small, usually irregularly shaped plot. Clearing is incomplete, reflecting both the limited tools available to the farmer and evolved ecological wisdom; stumps and roots are left in the ground, thus minimizing disturbance of the soil, holding down erosion, and providing for rapid forest regeneration after the plot is abandoned. Typically, the newly cut vegetation is burned after clearing, adding mineral-rich ashes to the soil.

The agriculture practiced by shifting cultivators is characterized by diversity and complexity. Several different crops are planted at once, sometimes including more than twenty different species and sub-species. This "mixed cropping" spreads harvesting efforts over the year, provides security in case one crop variety fails, and, by providing a varied environment, reduces the chance of wholesale losses to pests or disease. It also means that the soil is almost continuously protected from the destructive influences of sun and rain.

With both the soil's initial limited fertility and the ashes of burned vegetation to draw upon, crops produce fruitfully in the first season, but yields decline rapidly thereafter. Soil fertility is quickly depleted, profuse weeds that compete with crops for nutrients and sunlight invade the plot, and predatory insects multiply. Harvests may fall by 20 to 50 percent after one year, and within two to five years after clearing the plot is abandoned to nature. Grasses and weeds, then shrubs, and eventually, if the soil is left idle long enough, forest trees reclaim their right to the land and the soil builds back to its former state.

Plots are ideally left to rest for from six to over twenty years, depending on local conditions, before the cycle of clearing and desertion is repeated. Historically, many clans or tribes in the forest have led nomadic lives, moving their homes as they cleared new plots. Today, farmers more commonly live in permanent settlements and rotate cropping on adjacent lands. Throughout most humid and semi-arid zones in Africa, where some form of shifting cultivation remains the dominant means of food production, the land is never given time to return to high forest. Instead it returns only to the shrubbery stage; hence the usual labeling of African farming as a "bush fallow" system.

Shifting cultivation is often seen as a stigma of backwardness, but where the ratio of land to people is high enough, it is an ecologically sound system and one well adapted to tropical forest conditions. Few of the men and women who practice it have been educated in primary schools, let alone universities. Yet they usually carry in their heads an extraordinary fund of scientifically sound knowledge about plant species and soil qualities. Despite its ecological soundness, however, shifting cultivation has serious disadvantages. Production may be fairly secure, but the system makes no provision for the accumulation of investments that is the key to higher productivity and wealth. Most societies try to mold the environment in directions that serve human ends, but shifting cultivation implies the almost complete molding of human practices to fit the environment as it is. Material poverty is the inescapable condition of the traditional shifting cultivator.

Most importantly, in many regions the system can support only an extremely sparse population. Though conditions vary widely, often at least fifteen hectares of land must be available for each dependent person if shifting cultivation is to be sustainable over time. Once a threshold point is reached at which farmers return to a plot before its fertility is fully restored, a dangerous cycle of degeneration begins, unless skillful and painstaking measures to recycle nutrients and organic matter to the soil are scrupulously carried out.

In most agricultural regions the principal constraint on a rapid boost in food output is the slowness with which farmers adopt improved technologies well known to scientists. Inflexible or oppressive political, social, and economic institutions often frustrate the quick implementation of research results, as does the understandable hesitation of most farmers to risk substituting the unknown for the reliable. As a subject of agricultural research, however, the humid tropics are a giant step behind other zones. For the unpleasant truth is that, for many tropical areas of Africa, Latin America, and Southeast Asia, no alternative food-production system to shifting cultivation has yet proven both biologically and economically workable.

The current plight is in part a product of the colonial era, when tropical agricultural research concentrated heavily on improving plantation crops like cocoa, bananas, and rubber, and other crops grown mainly for export, such as cotton and peanuts. It is also due to the singular environmental rules of the humid tropics, which differ so greatly from those under which modern temperate-zone agriculture developed that farming systems cannot simply be transferred from one zone to another. Clearing the land cleanly, plowing the soil, and planting a single crop in neatly divided rows all severely aggravate erosion, deterioration of soil structure, leaching of nutrients, and pest damage.

In the former Congo (now Zaire), Belgian colonists tried strenuously to devise a system of continuous agricultural production to replace shifting cultivation—and failed. In their initial experiments in the 1930s, Belgian agronomists attempted to duplicate European intensive cultural practices. They cleared, burned, and uprooted the forest; plowed the soil, sowed a green manure crop (legumes or grasses); plowed again; sowed a two-year rotation of rice, manioc, or peanuts; and, finally, sowed another leguminous cover crop. The results were disastrous. Deep plowing damaged the soil structure; clean weeding and planting in rows led to accelerated soil erosion and leaching of nutrients, and left the soil exposed to the sun for thirteen out of twenty-four months, which in turn burned away its essential organic components. The cover crops had few of the beneficial effects on soil fertility to which European farmers were accustomed. René Dumont has described the cumulative effect: "The yield of every crop fell rapidly. The ears of rice did not swell. The 150 acres on which the experiment had been conducted had to be abandoned after a few years. Too far from the edge of the forest to be reseeded quickly and with the microclimate of its soil reduced almost to aridity, the area was still only thinly wooded ten years later."[3]

In response to this predicament, the corridor system, founded on the principals of the indigenous agriculture, was developed. In many ways it preserved the ecologically sound features of the traditional shifting cultivation, and yet managed to raise total production by allowing the systematic use of available labor, bringing more land under production, and facilitating technical assistance to Africans from Belgian agricultural extension agents. According to William Allan, the widespread development of corridor farming took place largely at the insistence of cotton companies, which saw its main purpose as increasing the acreage and yields of cotton.[4]

Several variants of the corridor system were adapted to different climatic, topographic, and vegetational conditions. A typical arrangement in a forest zone began with a large block of land divided into hundred-meter-wide strips, usually east-west oriented to take maximum advantage of sunlight. The number of strips in a block was dictated by the number of years required to complete a cycle of cultivation and regenerative fallow, usually something in the order of twenty years. Each strip was divided into a number of rectangular fields, which were then allocated to farming families, and every year another strip was cleared and cultivated. The period of cultivation for the strip usually ran two to four years, and then was followed by a lengthy fallow period. Thus a constant number of fields were kept in cultivation while the farmers moved slowly through the block. Cleared strips were alternated with uncleared ones so that fields were closely bordered by forest to shade the crops and provide seeds for the rejuvenation of natural vegetation during the fallow period.

Following Zaire's independence in the early 1960s, the corridor system largely disappeared. Though a vast improvement over efforts to duplicate Western-style mechanized agriculture, its passage into the history books has not been widely lamented. The corridor system was a creative adaptation of traditional methods to the requirements of the colonizers, but from the tropical ecological perspective it was no great improvement. The straight lines and rigid timetables for crop rotation satisfied a European need for order, and facilitated the harnessing of African labor for Belgian ends, but meant a loss in the flexibility so well-suited to the tropical environment. In any case, the corridor system, like the less mathematically precise shifting cultivation it replaced, was still an *extensive* use of the land that required large empty areas to sustain a relatively small annual product.

Technological adaptations to increase the land's carrying capacity have been a constant historical theme both in and out of the tropics as rising populations put pressure on the land. It is only within the last few centuries that shifting cultivation has disappeared in Europe, to be replaced by the integration of crop rotations, animals and their manures, and, eventually, chemical fertilizers into the food-producing system. In Southeast Asia, an intensive, ecologically stable system of irrigated rice production has evolved over the centuries. In Africa, too, cultures and technologies have adapted to population pressures in various ways. Where the tsetse fly does not limit cattle herds, as in some of the savannas of West Africa and the highlands of East Africa, for example, the careful use of animal manures has permitted nearly continuous cultivation without a great loss of fertility over time.

Other regions have been less fortunate. When populations grow while the amount of land available to a tribe or clan remains stable, the regenerative period allowed a given area between plantings must be constantly shortened. In some areas, as the soil fertility is gradually sapped, a tenacious grass takes over that is almost impossible to eradicate with traditional implements. Whether or not this occurs, farmers are forced to clear larger plots and to move more frequently as yields shrink. Reduced fertility means that crops produce less total organic matter and, subsequently, return less to the soil to guard against erosion and to rebuild soil structure. The deterioration of the land thus gathers momentum as topsoil washes away, lessening— sometimes permanently, more often for a long period—its life-supporting capacity.

The phenomenon is already prevalent enough in Africa for William Allan, in his classic work, *The African Husbandman,* to devote a chapter to Africa's "cycle of land degeneration." The reduced fallows of which Allan speaks have probably wreaked their most baneful destruction in the heavily settled drier areas of the West African Sudanian zone, such as northern Ghana and Nigeria, where they are partially responsible for destructive dust storms, and in parts of East Africa and Ethiopia.[5] But the cycle of degeneration is also a mounting threat in many humid tropical areas. The existence of vast unoccupied spaces and low national population densities can promote a false security. Where traditional technologies persist—due either to the lack of proven alternatives or to the failure to reach farmers with new methods—deterioration of the land begins as soon as human numbers in a local area surpass the level shifting cultivation can support. Africa's post-war population explosion, accompanied by the general perpetuation of traditional farming methods, means that the problems of soil erosion and declining fertility could accelerate dramatically throughout much of the continent over the coming decades.

The bellwether for many humid tropical regions may be eastern Nigeria, the most densely settled region south of the Sahara. Nigeria's former Eastern Region received world attention under the name of Biafra when it unsuccessfully tried to break away from the country in the late 1960s. But before that civil war, Barry Floyd, a geographer, wrote:

> The Eastern Region of Nigeria has one unfortunate claim to fame which might better be left unpublicized, except that it relates to a problem of monumental proportions. . . . Within the region there occur some of the most spectacular examples of soil erosion and "badland" topography to be

seen in West Africa. Erosion gullies attain a degree of severity and destructiveness seldom experienced in other parts of the continent. . . . Soil deterioration and degradation, in terms of the progressive loss of nutrients and breakdown of structure, is well nigh universal, due largely to over-farming and primitive, destructive methods of cultivation.[6]

Shifting cultivation has given way to virtually continuous cropping in parts of this region, undermining the soil's fertility and ability to resist erosion in an area already geologically susceptible to severe erosion.

Donald Vermeer has provided a pertinent description of cropping changes over time among the Tiv people, who live north of the region described above.[7] While the average population density of the Tiv tribal area was thirty-five per square kilometer in the early sixties, in the better farming regions it exceeded 230 per square kilometer and approached the higher densities to the south. Wherever the population density surpassed about 150 per square kilometer, Vermeer found, the traditional cropping system was no longer sustainable, and farmers began changing their practices to cope with the deteriorating situation. The Tiv population now is in the process of doubling in twenty-three years, a trend that compels great and rapid changes in the tribe's way of life.

Yams, the most highly prized food of the Tiv, have traditionally been the first crop planted, with great ceremony, on freshly opened land. Yams make heavy demands on soil fertility, and this crop pattern has provided the maximum possible yield. In the second year, millet or corn is customarily planted, with sesame as a follow-up cash crop. Occasionally, crops of peanuts, sweet potatoes, upland rice, or cassava are then raised in a fourth, fifth, or even a sixth year. Finally, the land is left idle for ten or more years to return to grasses and, preferably, woody vegetation.

In the southern areas of Tivland, mounting population density is forcing radical alterations in these practices. More and more fields are cropped for five or six years and left to regenerate for only one or two. As a result, writes Vermeer, "soils in the densely populated areas are becoming increasingly impoverished and degraded, a condition manifest by increasing extent and dominance of *Imperata* grass, soil erosion, and exposure of lateritic outcrops and loss of crop yields." The fertility problem is exacerbated by the growing shortage of fuel wood in southern Tivland, where the former forest has almost totally given way to farms and grass-lands. The stalks of harvested grain plants, historically left on the ground to add nutrients and organic matter, are now commonly uprooted and burned for cooking.

As fallow periods decline, the cropping sequence is changing. The primary role of the yam in the Tiv culture is being undermined. Soil fertility has fallen to such an extent that planting yams first no longer guarantees a good harvest; instead, a quickly-maturing food crop of sweet potatoes or peanuts often now initiates the cycle, followed by corn or millet in the same year. Yams are then planted in the second year of the cycle. Since yam yields have declined, grains, which in the past were mainly used for beer production, and which served as a dietary staple for only a month or two out of the year, have become the principal food for over half the year. One compelling reason for the shift to a new quick-growing food crop at the beginning of the cropping cycle, Vermeer suggests, is the ominous emergence of a "hungry season" in southern Tivland during the late months of the dry season each year. He concludes that "without alteration of cul-tural values and practices, continued deterioration of the environ-ment may lead to further reduction in crop yield, intensification of the hungry season, and possible triggering of a host of problems associated with under-nutrition."

Most tropical forest areas of Central and West Africa are far less heavily settled than eastern Nigeria, but this is no ground for compla-cency. The particular blend of techniques and soils of one region may safely support far fewer people than that of another, and, even if more productive agricultural methods are available, there is no assurance that they will reach the farmers in endangered zones, or that farmers will quickly adopt them. Furthermore, populations in localized areas may swell beyond ecologically safe levels even while nearly empty lands lie nearby; Nigeria provides one such example. African farmers squeezed out of crowded homelands are often, and understandably, more inclined to migrate to the cities than to take up farming in areas that are environmentally and socially unfamiliar or even hostile.

Thus reduced fallow cycles may already be damaging soils over rather substantial areas in Africa. Unfortunately, no one really knows how widespread the problem is; even national government officials in

Africa usually have only undocumented impressions of long-term changes in yields and soil fertility in the less modernized regions of their countries. At the seventh FAO Regional Conference for Africa, held in Gabon in September, 1972, delegates from several countries expressed concern about deteriorating soil fertility and quality due to reduced fallow cycles, provoking a special 1973 FAO Seminar on Shifting Cultivation and Soil Conservation in Africa. Estimates of the proportions of the problem were noticeably lacking at the seminar. Martin Billings, an agricultural expert with the U.S. Agency for International Development, suggests that crop yields are declining "in wide areas of Volta, the Guinea Coast, Zambia, Malawi, and Madagascar."[8]

Reliable facts on the extent of Africa's cycle of land degeneration are hard to come by. What we do know is that per capita food production in Africa has declined over the last twenty years—a distinction shared by no other continent.[9]

One continent to the west, the largest continuous forest in the world, which covers five-sixths of the basin of the Amazon River system in South America, also faces ecological stresses, though from causes quite different than those affecting eastern Nigeria. Larger in area than Europe (excluding the Soviet Union), and cradling a river that provides one-seventh of all the water flowing into the earth's oceans from land, the Amazon Basin is more sparsely populated than any region on earth, save the frozen wastes of the polar zones and the driest deserts. Outside the region's two small cities, the average density is less than one person per square kilometer of land.

In few areas has the imprint of human beings been so inconspicuous; the small indigenous Indian population lives by shifting cultivation in widely dispersed semi-nomadic settlements. Today, however, the governments that control the Basin—including Brazil, which occupies the bulk of it, as well as Venezuela, Colombia, Ecuador, Peru, and Bolivia—are doing their best to establish a sizable and lasting modern presence. Ambitious new roads are being cut through the forest, opening the way for colonists from the land-short Andes and from drought-prone northeastern Brazil, where large sugar estates co-exist uneasily with a large landless class. Concessions to utilize huge blocks of the Amazon Basin for forestry or ranching have

been granted to entrepreneurs and international corporations. Unex-
amined geologic formations are luring in prospectors for minerals and
oil, and occasionally new mining operations follow.

Governments are pressing into the Amazon Basin for various
reasons. Opening new farmlands will, it is hoped, defuse demands for
land reform and relieve the poverty of other regions. The geopolitical
urge to integrate wilderness areas into national political and economic
life before some other country does so is an undisguised motive. The
notion that gigantic windfall economic benefits await those who first
master the jungles cannot be discounted. These drives are capped by
a basic urge to conquer one of the world's last frontiers, an urge with
roots deeper than economic or ecological rationales alone.

By any account, the soils of most of the Amazon Basin are poor
and could best be exploited through forestry; some areas would best
be left alone in their natural state pending a better understanding of
the region's hydrological, climatological, biological, and cultural systems. Only a small fraction of the basin can profitably be farmed for
food crops, given available technologies and foreseeable market conditions. The area is far from adequately surveyed, and estimates about
land quality vary according to assumptions about the technical profi-
ciency of farmers. Brazil's federal aerial photography agency has
recently calculated that only 4 percent of the country's Amazon soils
have medium-to-high fertility.[10] This is still, of course, highly significant, involving a total area nearly the size of Great Britain. But
whether very much of its potential productivity will be tapped depends on many factors, including the cost of transporting necessities
into the forest and products out of it, the comparative cost of increasing agricultural output elsewhere in Brazil, and whether or not farmers have the knowledge or desire to protect soil quality for sustained
use. Most of these better soils are in the *varzea,* narrow plains along
the banks of some major rivers that receive annual floods and deposits
of fertile silt. Large-scale agricultural exploitation of the *varzea,* however, will require tremendous expenditures for drainage, dikes, and
other water control facilities.

All the Amazonian governments have programs to help new
farmers from other regions settle in the forest. But the spontaneous
influx of the land-hungry has in most cases overridden careful plans.
Since 1971, when a major stretch of Brazil's Trans-Amazon Highway

(which, when completed, will stretch from the Atlantic to the Peruvian border) was opened, about fifty thousand families have moved in to take over the hundred-hectare plots available at low cost along the edges of the new road. The total has been short of initial government hopes, but, with limited financial and administrative resources in the Amazon, and migrants who often know little about farming, let alone the particular demands of the humid tropics, the government has been forced to scale down its expectations markedly. In 1971, government financial aid for movement to the Amazon was stopped, since more people than the settlement agency could handle were making their way to the region on their own. As of 1975, the program to screen settlers outside the Amazon territory has also been discontinued, yet the brisk inflow of self-elected migrants continues.[11]

While a few of the more skilled colonists manage both to feed themselves and to produce fruits, cocoa, or pepper for sale, most of them engage in subsistence farming—producing only enough corn, manioc, sweet potatoes, and pineapples for home consumption. Given the general poverty of the soils and the absence of fertilizers and technical training, this is itself an accomplishment, if hardly the realization of the cornucopia many migrants dreamed of before they moved. What is not known is how many colonists will find they can't make a living in the jungle, and then sell their plots to more skillful neighbors or corporations; how much of the newly cleared earth will be severely degraded by over-intense, inappropriate cultivation; and how many of the colonists will be forced to abandon their plots after mining the soil of its fertility for several years. For all the emotional arguments the opening of the Amazon has generated, pro and con, surprisingly little analysis of the experience on the ground has yet emerged from the area. In any case, the colonization of the Amazon Basin is an experiment, and the final results will not be in for some time to come.

The major earlier effort to colonize the Amazon Basin does not offer an encouraging precedent. The government promoted settlement in the Bragantina region, east of the port city of Belém, in the late nineteenth century, and a three-hundred-kilometer railroad was built to join Belém and the town of Bragança, to the east. The initial wave of settlers, many of them from southern Europe, helped destroy the forest, but most soon drifted away, unable to cope with infertile soils and severe pest problems. Successive waves of migrants from northeastern Brazil and elsewhere cleared more land and practiced shifting cultivation for food crops in an effort to squeeze a living from the land. By the mid-twentieth century, the countryside held eight inhabitants per square kilometer, far more than are sustainable by shifting cultivation in this region.

Harald Sioli, a noted German scientist, recently described the results:

> Luxuriant high forest was transformed into extensive stretches of stunted scrub and only a few skeletons, now becoming rarer and rarer, of isolated jungle trees still testify to the former exuberant growth in the region. The nutrient storage of the former forest community, as well as the water-holding capacity of the soil have been upset, and local changes of the climate in form of longer droughts have been produced as a result of large bare areas. The final result of this effort at "development" was reached in a relatively short time. The introduction of fiber plants, especially Malva *(Pavonia malaco-phylla)* and Uacima *(Urena lobata)*, brought a short-term recovery in the general decline. But in general these new crops—as well as the use of the capoeiras [scrub vegetation that takes over the fallowed lands] for charcoal production—serve mainly to complete the process that produced a "ghost-landscape," as Eugenia Egler has called it, in less than 50 years.[12]

Most other colonization schemes in the lower Amazon have, like that in Bragantina, produced poverty-stricken people and landscapes, though nowhere else has destruction taken place on such a large scale.

More encouraging is the experience of Japanese immigrants to Amazonia, which has proved that at least some kinds of ongoing farming can be profitable. This group has concentrated on high-value crops well adapted to the tropical environment; two of their introductions, black pepper and jute, have become Amazonia's most valuable agricultural exports. The Japanese have successfully adapted a tradition of intensive soil husbandry to both *varzea* and upland areas of the Amazon Basin.[13]

Large-scale commercial enterprises, many of them foreign-owned, are likely to alter the Amazonian landscape far more visibly than colonists over the coming decades. Immense concessions of land totaling some seven million hectares had been granted Brazilian and foreign cattle-ranching companies by the end of 1973. One firm owned by D. K. Ludwig of New York has purchased 1.2 million hectares around the River Jari, just south of Surinam. The company is establishing diverse enterprises, including forestry, grazing, and paper, cellulose, and aluminum factories. It has built its own port facilities and thousands of miles of roads, and has plans for hydroelectric development and its own city of thirty-thousand inhabitants. Another U.S. firm, Georgia-Pacific, purchased 260,000 hectares of land in the mid-sixties and soon became the Amazon's largest producer of wood products. The famous King Ranch of Texas holds about seventy-three thousand hectares in Pará State and is clearing more than half of it for beef production.[14] The long-term ecological and economic viability of these schemes remains to be seen, as do the full economic costs and benefits accruing to the people of Brazil from this mode of exploiting the Amazon's resources.

Some observers fear that the combination of accelerated colonization by farmers using inappropriate techniques and wholesale forest clearance for quick profits through lumbering or grazing will create an ecological nightmare in the Amazon Basin. Overexposed soils could suffer an irreversible loss of fertility, and large-scale forest clearance will have unknown effects on the region's climate and river flows. Geographers at the Instituto Brasileiro de Geographia have warned that "a disaster of enormous proportions" is imminent. William Denevan, an American geographer, writes: "It is unlikely that the Amazon Basin will become desert, as sometimes claimed, but it could become a wasteland with greatly reduced opportunities for plants, animals, and man."[15] Others, noting that current Brazilian plans for settlement and commercial leases will cover little more than a fifth of the Basin's total area by the end of the century, are less noticeably worried. Whoever is right, it should be stressed that the debate today is not whether or not the Amazon Basin will soon become a major source of food for a hungry world, but whether or not the region will survive the coming onslaught of development without an ecological collapse, or at least a permanent cutback of productive capacity.

The challenge now shouldered by tropical agricultural researchers is the development of farming systems that permit continuous productive use of the land. For many regions, the knowledge needed to improve yields within the prevailing shifting-cultivation system is already available, and these gains have not yet been fully exploited. But scanning the next few decades, keeping in mind that populations in many tropical areas have been doubling in a little more than twenty years, reveals that the evolution of systems based on more intensive use of at least some humid tropical lands will be essential.

Many of the potential components of a sustainable continuous-cropping system are now known. These include zero-tillage techniques, which involve minimizing soil disturbance during planting and harvesting and leaving unharvested plant residues on the ground surface; and mulching, the systematic application of organic matter to the ground. Mixed cropping, in imitation of traditional farming, can further hold down soil degradation as well as pests. The most intractable challenge may be the maintenance of soil fertility over time within the financial and technical constraints that bind the average tropical farmer. Intensive applications of chemical fertilizers and, where necessary, applications of lime to fight the common problem of soil acidity can, if expertly monitored, support permanent cropping. But as D. J. Greenland, research director at the International Institute for Tropical Agriculture, puts it, "The real question is not whether it is possible but whether it is economic."[16] Given the low cash value of food crops, it makes no sense for the average farmer to spend money on chemical fertilizers. Therefore, the next step for researchers is the development of cropping sequences involving legumes and other green manure crops, mulches, and animal wastes that can hold down the need for costly commercial fertilizer inputs, which would in many cases have to be imported from abroad. Acid-tolerant crop varieties are also sorely needed for some regions.

If the building blocks of intensive tropical agriculture are now within sight, they have not yet been pieced together into a sturdy, reliable structure. Progress in this research will not come quickly. Once possible new agricultural systems are identified, they must be tested over many years; then, tested systems must be painstakingly adapted to local social and ecological conditions. New institutions to reach farmers with information, inputs, and marketing facilities are also necessary. Finally, the people who actually do the farming must be shown that it is in their own best interest to adopt the new ways.

In some parts of Africa, Asia, and Latin America, the intensification of agriculture in the tropical forests appears unavoidable. We can only hope that appropriate scientific knowledge will materialize and spread before the devastation of land and cultures proceeds too far. In most humid tropical zones, however, less demanding and more certainly proven modes of land use make far better sense. Usually it is wiser to intensify food production in other zones and to exploit the tropical forest in ways more naturally suited to its ecological traits as through forestry, or through tropical tree crops like bananas, cocoa, or oil palms. Clearing the Amazon forests of Brazil, in particular, may emanate from many possible rationales, but food production is not one of them. The agricultural gains possible through plowing new lands, redistributing large holdings, and improving cultivation practices outside the Amazon jungles are far more significant and technically feasible.

Tropical forests are often improvidently cleared for farming when, in fact, the land would provide greater long-term returns through forestry. Forests can provide wood products, fuel, protein in the form of wild game for nearby residents, and employment—not to mention their role as ecological stabilizers. Food production can instead be stressed where the best soils lie. In alluvial zones of Africa and Latin America, for example, it is often possible to adopt the extremely productive wet-rice cultivation of Asia. The high capital investments required and the unfamiliarity of techniques, rather than inherent environmental differences, seem to be the main factors constraining the spread of intensive rice cultivation.[17]

A promising hybrid of forestry and farming in the tropics is the *taungya* system, of Burmese origin, which is receiving increasing attention among tropical foresters and agronomists.[18] Also known as agri-silviculture, it provides an ecologically sound way to raise productivity and incomes in the tropical forest. Dozens of variations are possible, but the basic concept involves, first, the clearing of land and growth of a cycle of food crops. Then, before the last crops are harvested, fast-growing trees are planted among them. Next the land is given over to a forest plantation, providing wood products and employment. If the species are properly selected to allow sufficient recuperation of the soil, another cycle of food crops is possible after the wood is harvested. The *taungya* system retains the ecological soundness of well-managed shifting cultivation, but allows continuous productive use of the land rather than use limited to just two or three out of ten or twenty years.

Some ecologists view any agricultural intensification efforts in the humid tropics as futile and dangerous. Unfortunately, however, the only alternative to development looming ahead for many areas is not a pristine forest lightly touched by thinly spread shifting cultivators, but forests ravaged by people hungry for land and food, and a consummate deterioration of soil quality as the ratio between fallow periods and cultivated periods drops.

Given the ecological, cultural, and economic facts of tropical development, these lands will probably never be breadbaskets for the world. The people of the humid tropics will be fortunate if they prove able to feed themselves in the coming decades.

9. Dual Threat to World Fisheries

Harvesting food from the oceans is quite a different operation from coaxing sustenance from the land, but in water, as on land, the effort to squeeze too much from a natural system results in a harvest of less rather than more. Overfishing, like overgrazing, upsets the natural balance. And the mounting tide of pollutants dumped daily into the oceans is a biological time bomb of unknown megatonnage, with a fuse of indeterminate length, that jeopardizes the very existence of marine life.

Until quite recently, the oceans have seemed nearly boundless and endlessly bountiful. Fish have almost always been there more or less for the taking; the size of the catch was dependent only on the fisherman's efforts and skills. Early fishermen were restricted to inland ponds, rivers, and ocean coastlines. First on logs, later on rafts, and finally on boats, their ability to capture fish away from the shorelines grew, but the problem of spoilage still held most fishing vessels to within several hours of population centers. The nineteenth-century introduction of ice storage on fishing boats greatly expanded their range, permitting cruises of two to three weeks. By the mid-twentieth century, massive fleets and floating factories able to process and freeze fish right on board could make journeys of many months and, with the aid of helicopters and sonar, could locate and chase down schools of fish almost anywhere.

In the face of improved fishing capacities and a soaring world demand for protein, the oceans have been generous. Between 1950 and 1970 the world fish catch more than tripled; rising from twenty-one million to seventy million metric tons, it reached a new record almost every year. The global catch grew at over 5 percent annually, or more than twice as fast as world population, thus significantly augmenting the per capita supply of marine protein. The 1970 world catch averaged out, in live weight, to nearly forty pounds per person.

Fish are an important protein source in many of the world's more populous countries, including Japan, where the average person consumes seventy-one pounds per year; the Philippines, where consumption averages forty-four pounds; and the Soviet Union, where twenty-three pounds is the average. These figures do not include the third of the world catch that is ground into feed and eaten almost entirely by poultry and hogs in Japan, Europe, and the United States. Though the richer countries catch or buy a disproportionate share of the total, fish are virtually the only source of high-quality animal protein for hundreds of millions of the world's poorest people in parts of China and Southeast Asia, and elsewhere in Africa, Asia, and Latin America.

In the mid-seventies, as the era of rapid growth in world fisheries draws to a close, difficult questions are coming to the fore: does human society possess the organizational capacity to preserve this valuable, renewable resource at a high level of productivity? What criteria will be used to divide up the finite ocean harvest when the rich and the poor, the overfed and the malnourished, the powerful and the weak all clamor for more animal protein? The continuing United Nations Law of the Sea Conference marks a transitional period to a new order in the oceans. It remains uncertain whether this will be an order of international cooperation or conflict, of enlightened fisheries exploitation or myopic resource destruction.

How much longer can we expect the world catch to grow? Food and Agriculture Organization scientists, in collaboration with fishery scientists and institutions throughout the world, recently estimated a maximum sustainable world catch of roughly one hundred million metric tons annually. This total includes all species currently sold in the world's fish markets, ranging in size from whales to anchovies, and also species not yet exploited commercially but similar in size and appearance to those that are. It assumes that currently underfished regions like the Indian Ocean, the South China Sea, and others will be exploited; that few major stocks will be seriously depleted by overfishing; and that pollution will not destroy marine life over significant areas. Not taken into account is the large-scale utilization of new species low on the food chain, such as the Antarctic krill, a prolific plankton with harvest possibilities of up to fifty million tons should the problems of cost and palatability that now limit its exploitation be overcome. Nor does the estimate include large potential production increases through aquaculture, or fish farming.[1]

The possibility of raising the annual global catch from its current seventy million tons to one hundred million tons seems, at first glance, an encouraging conclusion—until one runs through the mathematics of growth. If the rate of increase in the world catch of the 1950–70 period could be maintained, the estimated sustainable limit of one hundred million tons would be reached in just *eight years.* Even if the FAO estimate turns out to be overly pessimistic, and the catch could be doubled to 140 million tons, as some feel is possible, this higher limit would be reached in just *fourteen years* at the accustomed growth rate. Every major fishing nation plans to increase its catch, and fleets expand every year, but it remains an open question whether or not the international power to limit fishing efforts can be developed before the potential additions to the catch are used up.

An overextended world fleet could rush past the safe limit quickly, severely damaging many species and causing a partial collapse of the oceanic harvest. Or, perhaps more likely, the growth in catch may now slow down markedly despite intensified fishing efforts, as the catch of the more easily captured stocks peaks and, in some instances, declines. In either case, the world will probably not experience a sustained, dramatic increase in the supply of marine protein again. A gap in the annual increment of world food production is emerging that can be filled only by land-based protein sources.

By the early seventies, serious stresses in the oceanic system had already emerged, raising doubts in some quarters as to whether the catch will ever reach even one hundred million tons. Two decades of continual growth came to an abrupt halt in 1971, when the catch fell to just below the previous year's total of seventy million tons. Then, in both 1972 and 1973, as the important Peruvian anchovy fishery temporarily collapsed, the total world catch dropped still further to under sixty-six million tons. The preliminary estimate for 1974 shows a catch back up to the 1970 level, or between sixty-nine million and seventy million tons.[2]

This sharp dip in the supply of marine protein caught most analysts off guard. The glare of spectacular growth helped mask the less prudent aspects of the process by which that growth was achieved. Hidden beneath the overall growth figures was a continual movement to new species and regions as traditionally preferred species were fished to minimum replacement levels—or beyond. Centuries of unregulated growth in world fisheries have taken many casualties, and such cavalier abuse of the oceans' life cycles was bound to catch up with us one day. A few species have now been virtually wiped out altogether, some have been depleted to a fraction of their potential yield, and many are producing at below their maximum level due to overly intense fishing pressures.[3]

The process by which some fisheries have grown, suffered overexploitation, and then declined is charted well by the experience of the Northwestern Atlantic offshore region extending from Rhode Island northward to the southern coast of Greenland. Fished commercially for 350 years, and accounting for 5 percent of the total world catch, this region is in some ways a microcosm of world fisheries and a harbinger of things to come elsewhere.

The northwest Atlantic is biologically rich and, as the Soviet and European fleets joined those of Canada and the United States, the area's fish harvest rose steadily from 1.8 million tons in 1954 to a peak of 3.9 million tons in 1968. But though the level of fishing effort has grown with each succeeding year, the total catch has not surpassed 3.5 million tons in any of the five following years for which data are available.[4]

The falling catch of several of the region's most sought-after species is largely due to overfishing and explains the drop in total output. The harvest of cod, the most important single species, reached a high of 1.9 million tons in 1968; by 1973, it had fallen by more than 50 percent. The catch of herring also peaked in 1968, at 0.9 million tons, and was down to 0.5 million tons in 1973. The haddock catch turned the corner earlier, in 1965, when the catch reached 249,000 tons; it had fallen to little more than a tenth of that level by 1973.

The nations fishing in the area have long maintained an International Commission for the Northwest Atlantic Fisheries to collectively oversee the fishery. Though the commission has not been able to reduce fishing efforts to a safe and economically sensible level, its recent strenuous efforts to reduce the pressure on the more gravely threatened species in various sub-regions have, in some cases, helped prevent their total depletion.

In the southeast Pacific Ocean, off the Peruvian coast, the world's richest fishing ground has just experienced a convulsion that parallels strikingly such terrestrial ecological breakdowns as the recent Sahelian debacle and the American Dust Bowl of the thirties. In all three cases, humans made progressively greater demands on the ecosystem, pressing it to its limits. Then, when a sudden but predictable change in natural conditions occurred that temporarily reduced the system's capacity to support life, the confluence of pressures imperiled both humans and the natural system.

This particular tale originates in the mid-fifties, when the fishing industry of Peru began the most spectacular expansion in the history of marine fisheries. Ocean conditions off the coast of Peru and northern Chile facilitate an extraordinary concentration of life there. A westward swing of the Humboldt Current, which flows toward Peru out of the South Pacific, causes a cool, nutrient-rich upwelling from the depths, and, as it reaches sunlight, there is a rich bloom of plants and animals.

In the nineteenth and early twentieth centuries, humans capitalized upon this ecosystem by collecting for sale as fertilizer the manure (guano) of sea birds that fed upon the billions of anchovies swarming off the coast. Governmental regulations in reaction to the foreseeable exhaustion of the guano, together with the spread of chemical fertilizers, helped dampen this trade. But a new industry, based one step lower on the food chain, emerged to change the face of oceanic fisheries. At a time when technological progress permitted more efficient culling of the anchovies, rising incomes and a revolution in the poultry industries of Europe, Japan, and North America were creating a lucrative new market for the fishmeal into which the anchovies are ground. Unfortunately, the anchovies, a vast source of high-quality protein, have been mainly used not to eliminate malnutrition in the Andes and elsewhere in Latin America, but to help satisfy the growing taste for meat in the industrial countries. In part because of the prolific anchovy, fishmeal has become a frequent protein additive to feeds for broiler chickens that are often now produced in huge, scientifically fed concentrations.

The anchovies soon proved a precious natural resource for Peru, with fishmeal replacing even copper as the country's top export in the mid-sixties. As the catch soared, Peru became the world's leading fishing nation; its anchovy catch accounted for a fifth of the entire world fish catch in many years. Often, nearly two-thirds of the world's fishmeal exports came from this one country.

High profits generated massive over investments in fishing vessels and fishmeal-processing plants. Gerald Paulik described the staggering capacity of the Peruvian fleet in 1971: "On April 28, 1970, the total catch reached 9.5 million metric tons. The fishery continued for ten more days, taking 100,000 metric tons per day. This fantastic catching power could have taken the whole U.S. yellowfin tuna catch in one day, or the entire U.S. catch of all Pacific salmon in two and a half days. Obviously, this type of destructive power must be handled most carefully, and precise management and regulation are essential."[5]

Though an excellent system of biological monitoring and fishing regulation was developed, the combination of too many investors with too many boats and the government's hunger for foreign exchange generated strong pressures to continually increase the anchovy harvest. In 1970, an international team of biologists estimated the maximum sustainable yield of the Peruvian anchovy fishery at about 9.5 million metric tons. Political realities had stronger exponents than biological needs, however, and the catches in 1967, 1968, 1970, and 1971 all exceeded this estimate. The catch reached a height of over twelve million tons in 1970 and surpassed ten million in 1971.

The anchovy stock had long been known for its instability; an unpredictable warm current, locally known as *el niño* ("the child," because it often occurs around Christmastime), periodically drives the anchovies deep in search of colder water, and possibly reduces their numbers. Beyond this, the earlier collapses of fisheries based on species closely related to the anchovy—the sardines off California and herring off Norway and Japan—suggested to thoughtful observers the greater susceptibility of these types than most to spectacular, rapid damage.

In mid-1972, after heavy fishing in the spring, it became apparent that the anchovies had vanished from their usual coastal haunts in the wake of *el niño*. Alarmed by their absence, the Peruvian government halted fishing for the rest of the year. In early 1973, the characteristics of the current remained unusual, and experimental catches indicated the absence of a new brood of anchovies. Fearful that the fishery would be permanently destroyed, the Peruvians closed it for most of that year and part of the spring of 1974. For almost two years, the world's greatest fishery lay virtually idle—two very long years for the fishermen, who were out of work; for the industry owners, some of whom lost fortunes; for the Peruvian government, which lost its chief export; and for consumers everywhere, who found the world protein market disrupted by the sudden shortage of fishmeal.

There has been a tendency in many quarters to ascribe this collapse entirely to nature and *el niño*, but a careful examination of events makes that explanation unsatisfactory. In hindsight, observes John Gulland, a leading FAO fisheries expert, it appears likely that the overfishing of 1970 and 1971 depleted the stock to such a vulnerable condition that, "while recruitment [successful replenishment by a new spawning] could be maintained under average or better environmental conditions, the unfavorable conditions of *el niño* would cause a recruitment failure that would not have occurred if the adult stock had been larger."[6] After the particularly heavy catch of 1970, an exceptionally small brood was produced in 1971—well before *el niño* struck. The fishing effort of early 1972 apparently finished off the survivors of both the 1970 and low 1971 spawnings, leaving few adults to replenish the species. Tellingly, *el niño* caused only a brief and modest drop in the anchovy catch in 1965, after annual catches under the recommended level, and did not disrupt the following year's brood.

In 1973, the Peruvian government nationalized the fishmeal industry as part of an effort to reduce its overcapacity. The number of anchovy fishing boats was cut in half. Under the guidance of marine biologists, the government also tightened management of fishing levels. In 1974, as the fishery began to regenerate, a catch of just 3.6

million tons was allowed, but in 1975, as unstable conditions continued, the catch was smaller yet. With the new degree of management, biologists hope the fishery will be back to normal by 1977; presumably, fishing will henceforth be held to a sustainable level.

A source of food for humans, the oceans also serve as the ultimate waste receptacle for the planet. Deliberately and accidentally, the human race is adding to the oceans thousands of waste products—some highly toxic. Oil, chemical effluents, lethal chemical-warfare gases, radioactive wastes, junk metal, trace elements, organic wastes from humans and animals, automobile exhaust products, pesticides, and detergents are concentrated precisely in the small portion of the seas supporting nearly all marine life: along the continental coasts.

The long-term biological effects of polluting the ocean with industrial, military, municipal, and agricultural wastes are not yet fully known. Both the quantity and the variety of oceanic pollutants are growing faster than our ability to collect information about them and about their individual and synergistic consequences for the marine biosphere. The natural instability of many fish populations sometimes makes it difficult to tell whether overfishing, pollution, or a natural environmental change is responsible for a stock's collapse. It is certain, however, that oceanic pollution is now global in scale, and that it poses a serious threat to marine food resources.

Pollution of inland and offshore waters has killed some fish outright, jeopardized the existence of others by upsetting the ecology of spawning grounds, and rendered still others inedible.[7] In the United States, thirty-three states have banned commercial fishing in some freshwater bodies during the past few years because of dangerous traces of mercury in fish. Illness and death have been blamed on mercury-contaminated fish in Japan, and in Sweden the sale of fish from a number of inland waters and Baltic coastal areas has been banned because of mercury pollution. California mackerel fishing has been stopped because the concentration of DDT residues exceeds the acceptable tolerance level of five parts per million.

Fish have virtually disappeared from some of the more polluted rivers and coastal zones of the industrial countries. In the United States, over half the human population and 40 percent of all manufacturing plants are adjacent to estuaries and coastal waters. City sewage, much of which is untreated, is the major pollutant, followed by industrial effluents and agricultural chemicals. The once-rich oyster beds of Raritan Bay, New Jersey, are now almost devoid of this mollusk because of municipal and industrial wastes. Shrimp operations along the Gulf of Mexico have frequently been damaged; the harvest of Galveston Bay, Texas, for example, shrank by more than half between 1962 and 1966.

The litany of pollution-damaged fish stocks could continue for many pages. The combined effects of sewage, industrial wastes, heavy pesticide and fertilizer runoff, and oil spills have been increasingly visible over the last few decades to those living along many parts of the Mediterranean coast. Coastal fisheries often have suffered accordingly. Pollution has seriously damaged the valuable sturgeon industry of the Caspian Sea. Many of Japan's once rich fisheries, such as Tokyo Bay, Osaka Bay, and Hiroshima Bay, are now "dead seas," and the Inland Sea may soon follow.

Fishery damage by pollution is not limited to the heavily industrialized regions. The extensive use of DDT and other pesticides for malaria control and crop protection often kills fish in the poor countries, and wastes from mushrooming cities and industries are also a threat to lakes, rivers, and coastal zones. Anxious to build an industrial sector, few poor countries impose strict effluent controls on manufacturers, a safeguard gradually being adopted in the more affluent countries. Fish production in Egypt's Lake Mariut dropped from ten thousand tons in 1961 to below two thousand tons in 1967, due mainly to the growing load of sewage and industrial waste it receives. Along parts of the eastern coast of Mexico, the production of shellfish and bass has been undercut by wastes from sugar and pulp and paper factories; pesticide-related fish kills have been reported in Mexico and Cuba. Some ecologists fear that agricultural and industrial pollution threatens the productivity of the important Great Rift lakes of East Africa.[8] The problem there is plainly international, since these lakes provide food for people in eight bordering countries.

One of the bright spots in world fishery prospects is the apparent potential for radically escalating the practice of aquaculture, particularly in the protein-short poor countries. The cultivation of fish or shellfish in ponds and coastal lagoons is already widespread in Asia, but can most likely be multiplied severalfold worldwide over the next few decades. Aquaculture, however, is exceedingly vulnerable to pollution. Pesticide and fertilizer runoff has already reduced fish production in farm ponds in the Philippines, Malaysia, and Indonesia; this problem will certainly intensify throughout Southeast Asia and elsewhere as governments strive to boost agricultural yields.[9] Industrial wastes or petroleum washing up from sea lanes could render prime coastal aquaculture sites unusable. Coastal shellfish operations in the United States and Japan have already been destroyed by poisoned waters. The availability of habitable water may be the greatest single constraint on the growth of aquaculture.

Pesticides spread on the land find their way into estuaries, coastal waters, and the open seas; sometimes they are carried thousands of miles from the point of application by rivers, rain, and winds. As much as one-fourth of all the DDT produced so far has wound up in the ocean, and the world's fish are now almost universally contaminated.[10] Despite the banning of DDT in many industrial countries, its use worldwide, at about 150,000 metric tons annually, may now be higher than it was a decade ago.[11] The period of DDT's critical activity in the oceans is not known, nor are the effects on marine life of many of its substitutes. Their ready dissemination to the far corners of the earth makes the application of persistent pesticides anywhere a matter of global interest.

Even small amounts—no more than a few parts per billion parts of water—of DDT and many other pesticides can stunt or stop reproduction in some species. A concentration of eight parts per million in the ovaries of sea trout in an estuary off the Texas coast prevented spawning. As little as one part in ten billion of DDT in water severely cuts the growth rate of oysters, and two parts in ten billion kill commercial species of shrimp and crabs.

Over six million metric tons of petroleum enter the oceans each year, according to a recent study by the U.S. National Academy of Sciences.[12] This equals seven hundred tons every hour, and seventeen thousand tons every day. (Such an estimate can be little more than an informed guess; others range from five to ten million tons per year.) Some six hundred thousand tons of oil seep naturally into the oceans from shoreline and submarine oil deposits. Over the millenia, in fact, from fifty to one hundred times as much oil may have seeped naturally into the environment as is now left in known global petroleum reserves. Coastal refineries, urban and industrial waste disposal, and rivers carrying inland petroleum discharges together supply nearly half of all oceanic oil, while petroleum hydrocarbons in atmospheric fallout—a good share of which originate with automobile exhausts—rival natural seepage as the source of an additional tenth of the total.

Though dramatic accidental spills from damaged tankers or offshore wells often held the public eye, these presently account for less than a twentieth of the total. The quotidian operations of tankers transporting oil—loading, unloading, discharging ballast, and cleaning tanks—have presented a far greater hazard so far. Routine tanker operations and spills together account for more than a third of all oil entering the oceans, and only the widespread adoption over the last decade of improved tank-flushing techniques has kept this total from climbing much higher.

After eons of natural seepage and decades of increasing human-caused contamination by petroleum, marine life continues to thrive along most of the world's coastlines; apparently the oceans have generally been able to tolerate the influx at its current levels. As United Nations scientists point out, however, "the fact remains that once the recovery capacity of an environment is exceeded, deterioration can be rapid and catastrophic; and we do not know how much oil pollution the ocean can accept and still recover."[13] The critical point beyond which damage is massive and irreversible may be well above present pollution levels, or, like all sudden deaths, an ecosystem's collapse may take us by surprise.

When petroleum is discharged into the sea, parts of it evaporate into the air, parts begin biological disintegration as they are attacked by marine micro-organisms, parts fall as sediment to the ocean floor, and some parts disappear, their fate unknown to us. The heavier fraction of petroleum forms indestructible tar, floating lumps of which now blemish every ocean. Norwegian explorer Thor Heyerdahl's report, after his 1969 Atlantic expedition on the raft *Ra*, of "seemingly never-ending" clots of oil on the high seas has since been so often quoted as to become a cliché; yet clichés befit a situation that has become so commonplace. Oil slicks and drifting masses of tar abound throughout the Mediterranean. Tar regularly clogs the nets of biologists gathering animal specimens in the open Atlantic. Formerly unpolluted beaches on the east coast of Africa, in southern France, and on many islands in both the Indian and Atlantic oceans are increasingly defiled by tar masses washing up with the waves.[14]

Analysis of floating oil residue indicates that most of it comes from tanker operations, rather than from natural seeps or from the greater onshore pollution. As the transport of oil by sea continues to soar, the number of floating oil slicks and tar lumps will climb as well. If the ultimate impact of these drifting blobs and slicks on marine life cannot yet be calculated scientifically, the effect on human sensibilities is gut-level and immediate. Once oil washes onto coastal sands, its disintegration can take decades, and unless oil discharges can be held "to a level that can be assimilated through natural degradation processes," observe experts at the National Academy of Sciences, "we will all have to reconcile ourselves to oil-contaminated beaches."

Oil slicks, tarry lumps, and other unknown petroleum residues are already accumulating at a frightening pace, but the outlook for coming decades summons up an even more menacing picture. Two-thirds of all the world's oil production, or nearly two billion metric tons, will be transported by sea this year. By 1985, the amount of oil traveling the oceans may approach three and one-half to four billion metric tons.[15] If the ratio of pollution to total oil transported should remain the same, then the amount of oil intentionally and accidentally discharged into the oceans by tankers alone would nearly double to over four million tons annually.

Already reports of "minor" or "major" spills in some part of the world appear in metropolitan newspapers on an almost weekly basis. One typical day in early 1975, Robert Citron of the Smithsonian Institution sat before a U.S. Senate committee and described the week's headlines:

As we sit here today there are massive oil spills in Bantry Bay, Ireland, at Kurashiki, Japan, at Kilton Island near India, and off the coast of Singapore, where the *Showa Maru*, a 237,000-ton (deadweight) supertanker ran aground and dumped 3,300 tons of oil that spread and polluted the waters of Indonesia, Malaysia, and Singapore. A few days ago at St. Croix in the Virgin Islands, the harbor was inundated with 450 tons of oil; quite a bit of it was spilled in the harbor itself, some of it was spilled just outside the harbor. . . . Last night at 6:10 P.M. there was a collision of a freighter, *American Wheat*, and the tug *Mama Lere* on the Mississippi River near Chalmette, Louisiana. Three oil barges were released and went downriver in flames, and an estimated 10,000 tons of oil, or 2,700,000 gallons of oil, were spilled into the Mississippi. This occurred just 15 hours ago.[16]

International accords to curb oceanic pollution negotiated in 1972 and 1973 might, if ratified by enough countries, eventually help blunt the dangers of increased oil shipping. In addition to absolute limits on the international dumping of petroleum and other harmful materials into the seas, pollution-limiting structural features such as ballast tanks segregated from cargo tanks, and size limits on individual cargo tanks aboard ships, will be required of newly-built oil tankers. The adoption of ballast-flushing systems that minimize pollution will be required of most existing tankers.

If adopted and fully enforced (no easy task), these international conventions will reduce, though by no means eliminate, the chances of superspills from the giant new supertankers now on the drawing boards, and will hold down pollution from routine tanker operations. A surfeit of oil tankers in the mid-seventies has temporarily slowed the pace of supertanker construction; it also means that new ships incorporating more safety design features will likely enter maritime commerce more slowly. Petroleum in the oceans—from tankers as well as from the even more significant land-based sources—is sure to remain a major global ecological concern for the next few decades.

Concentrated local kills of fish and fowl from oil spills and industrial pollution are well known and increasingly frequent. What is not known is the combined impact of all the dissolved oil fractions, floating tars, pesticides, and industrial poisons flowing into—and, in some cases, accumulating in—the oceans. How will they affect the phytoplankton, the one-celled plants that are the foundation of the entire marine food chain? At some unknown date, the unrelenting river of foreign matter humans are pouring into the seas could measurably impair or halt altogether the life of phytoplankton over large areas. The fears occasionally voiced that the earth's oxygen supply would be threatened by the death of the phytoplankton are probably unjustified, as John Ryther of the Woods Hole Oceanographic Institution has pointed out.[17] But on this subject, perhaps *Supership* author Noël Mostert deserves the last word: "If the seas don't breathe, or if they breathe asthmatically and imperfectly, what else in our environment will struggle for breath? It is a simple-minded question that anyone rotating a globe two-thirds of which is marked as covered by salt water can't avoid asking himself."[18] Some fish stories don't need to be exaggerated.

10. The Politics of Soil Conservation

IN THEIR RELATIONS with nature, as in their relations with their fellows, human beings appear to be poor students of history. Despite the all too visible diorama of ruined landscapes and abandoned civilizations, mistakes of the past seem time and again to serve as models instead of usable lessons. Deaf to the almost daily warnings of some looming new ecological threat, governments and individuals rarely change their habits accordingly.

Studies of human-caused environmental degradation sometimes leave an aftertaste of near-misanthropy. How can people destroy the basis of their own survival? Given the record, it is tempting to blame the continuing devastation on human ignorance or stupidity. If the technical solutions to environmental degradation, nearly always known and publicized by scientists, were really just that—technical solutions—then this assessment of human knowledge and aptitudes might suffice as an explanation of the current environmental malaise. In some ways, at least, this would be an encouraging conclusion, suggesting the potential to solve these problems fairly easily and quickly through technocratic dictatorship.

For better or worse, however, the application of technical answers to ecological problems almost never turns out to be simple. Scientists and development planners work out elaborate schemes for rural regeneration, but peasants and goats seldom seem to find it in their own interest to assume the profile of the computer cards they are dealt. In this book, I have generally tried to place environmentally destructive habits in their historical and cultural contexts—to provide at least a sketch of the multiple structures and forces, ranging from land-tenure systems to the primal drive for survival, that undergird humans' treatment of land. Now it may be useful to discuss more systematically some of the many factors—the constellation of influences that must ultimately be termed political—that help propogate suicidal land use in the face of scientists' warnings.

Land-use patterns are an expression of deep political, economic, and cultural structures; they do not change overnight when an ecologist or forester sounds the alarm that a country is losing its resource base. In many countries, the deterioration of the land will not be halted until basic changes in land tenure and national economic priorities occur. This point is starkly illustrated by the story of an Ethiopian reforestation project related in Chapter 6.[1] Peasants whose personal prospects for progress were smothered by feudal land-tenure and social structures sabotaged a program to restore the land by planting tree seedlings upside down. The immediate cause of this non-violent rebellion was apparently the substandard (by local norms) level of wages the laborers were receiving. But, had the workers believed that an improvement in the land's quality would seriously improve their own welfare and that of their children, their behavior would almost certainly have been different.

In his study of El Salvador, Howard Daugherty documents other ways in which land-ownership and usage patterns can influence ecological trends.[2] El Salvador is one of the most environmentally devastated countries of the New World. The tropical deciduous forests that once covered 90 percent of the country have been totally destroyed by centuries of clearance for grazing, plantations, mining, charcoal manufacturing, and, especially within the last century, the spread of subsistence cultivation. Soil erosion has sapped fertility throughout much of the countryside, but above all on the extensive upland hillsides. A recent study by the Organization of American States concluded that 77 percent of El Salvador's land area is suffering from accelerated erosion.[3] Nowhere has the damage been more extensive than in the country's northern mountains, which were originally covered with dense forests. Long subjection to heavy cultivation and grazing has largely destroyed this range's topsoils, reducing many hills to a rocky, barren moonscape. While cultivation and grazing continue in this depleted zone, its reduced fertility has been a major cause of the heavy emigration of El Salvadorians into neighboring Honduras, which helped precipitate a war between the two countries in 1969.

The obvious catalyst of El Salvador's environmental decay in this century has been the dizzying growth in human numbers. The population of 775,000 in 1900 doubled by 1940, redoubled by the mid-1960s, and is currently in the process of doubling again within the space of just twenty-two years. El Salvador's population, the densest on the mainland of the Americas, guarantees heavy ecological pressures, especially since the majority wrest their living from the soil.

Behind the aggregate figures for population growth and land damage are some revealing statistics that more completely explain environmental trends. The pattern of land ownership is unquestionably a major contributor to ecological stress. *Latifundia*, large estates devoted to ranching and commercial crops, have been deeply entrenched in El Salvador since the Spanish occupation nearly five centuries ago. As of 1961, less than 1 percent of the country's farms were over one hundred hectares in size, but they occupied 48 percent of the agricultural land. At the other end of the scale, 47 percent of the farms were smaller than one hectare, and constituted less than 4 percent of the total farming area. The trend of recent decades has been a perpetuation of the large holdings and increasing fragmentation of the smaller farms.

The *latifundia* occupy El Salvador's most fertile and productive lands—the middle volcanic slopes, the interior basins and river valleys, and much of the coastal plain. Because of the export-market orientation of the *latifundistas*—the owners—a third of El Salvador's cropland is annually planted to coffee, cotton, and sugar cane. Meanwhile, hundreds of thousands of subsistence farmers, struggling to grow food for their families, are crowded onto the remaining land. All arable land has already been put to the plow, driving farmers onto slopes where erosion may require abandonment after a year or two of cultivation, and ensuring a reduction in the fallow periods essential for maintaining soil fertility and fighting erosion. Daugherty sums up the situation: "The increasing fragmentation of small plots and the progressive trend from shifting agriculture to semi-permanent and permanent cultivation (necessitated by the lack of additional land) has led to the widespread ecologic destruction of much of upland El Salvador, which comprises the bulk of the land surface of the country. Thus, the root of the ecologic problem of El Salvador is not only gross overpopulation, but the structure of land tenure as well, which has forced the intensive utilization of slope land."

If archaic, inequitable land distribution promotes suicidal uses of the land, maldistribution of economic opportunity also encourages the rapid population growth that itself has suicidal overtones. Poor, often malnourished subsistence farmers account for the greatest share of El Salvador's population growth, and, judging from experience elsewhere, they are unlikely to drop their preference for large families until they enjoy improved economic security and social conditions. Inequitable social structures, then, can help account for ecological deterioration both directly, by forcing undesirable land utilization, and indirectly, by fostering runaway population growth.

El Salvador's advanced state of physical degradation stands as a warning to many other countries—especially in Latin America, where the concentration of large landholdings in the hands of a few is generally far greater than in Asia and Africa—in which populations are multiplying fast and skewed land-tenure structures are encouraging the agrarian poor to destroy their own prospects for advancement. Most poor countries are not yet so densely peopled as El Salvador, but as the population doubles in two to three decades, acute pressures are sure to develop in localized, though sometimes quite large, areas.

Haiti is among the few countries that already rival or perhaps surpass El Salvador in nationwide environmental destruction. Not coincidentally, Haiti also resembles El Salvador in its inequitable distribution of land ownership, economic opportunities, and social services. It, too, has one of the highest birth rates in the Americas. Haiti, which means "Green Island" in the indigenous language, is now less than 9 percent wooded, and its mountains are completely ravaged. Long a consequence of severe poverty, soil erosion has reached the point where it is now a major *cause* of poverty as well. The UN Development Program has labeled "rapid and increasing erosion" as "the country's principal problem."[4] Wealthy farmers and North American sugar corporations own the best valley lands, crowding peasants onto slopes where cultivation is a futile, temporary proposition.

Steep slopes have been cleared by the land-hungry, and traditional fallow periods have been violated, in large areas of Asia, Africa, and Latin America. National population-density averages, which tend to disguise land problems, can be misleading; the exceptional sparseness of the Amazon Basin population, for example, provides little comfort to the landless peasants struggling to survive in northeastern Brazil. Often the redistribution of better lands would hold down the encroachment on marginal areas by farmers, desperate for a plot on which to grow food, who have been shunted off good farmlands by overcrowding and inequitable ownership patterns. A recent United Nations reforestation program near the town of Ayapel, Colombia, on formerly tree-covered slopes now suffering from severe erosion had to be abandoned because the project area was invaded by disenfranchised squatters.[5] The goal of ecological revival was sacrificed to the more immediate subsistence needs of the squatters, though in the long run both are life and death matters.

Without access to land in the more hospitable, productive zones, the landless may turn into vagabonds, farming wherever they can for brief spells until erosion, depleted fertility, or the police drive them on. Many give up and migrate to the cities; others move to tropical forest zones, such as the Amazon Basin and parts of West Africa, where large areas lie unused by people. After learning the hard way that jungle soils are not as easily mastered as those to which they are accustomed, farmers who had hoped to settle down to a more prosperous life find themselves and their families itinerants who are forced to destroy prime forests and soils as they try to carve a living out of the unfamiliar rain forests.

The influence of land tenure on cropping patterns, as illustrated by the case study of El Salvador, reveals other widely applicable lessons about ecologically sound land utilization. Cash crops for export like cotton, sugar cane, or beef are a key source of foreign exchange for many of the poorest countries, and can play a useful role in promoting economic development if the proceeds from their sale are properly spent. Seldom, however, is close attention given to what, from the perspectives of human welfare and ecology, an optimal mix of agricultural land uses might be in countries with shortages of good farmland and widespread malnutrition. Thus, in Costa Rica, the spread of cattle ranching to supply the North American market is forcing smaller farmers onto poor-quality, easily eroded lands, even as per capita beef consumption within Costa Rica drops.[6]

A narrow focus on aggregate GNP statistics and the desirability of extending the "modern economy" into backward areas may result in a heavy premium on commercial agriculture. But normal economic signals sometimes obscure the potential social and economic costs of large plantations or ranches squeezing subsistence farmers onto marginal lands, whether in drought-prone areas of Chad or on mountain slopes in Ecuador. Not only is the potential of these lands (perhaps for forestry, horticulture, grazing, tourism, or recreation) destroyed by intensive and improper cultivation, but huge costs in the form of floods, reservoir sedimentation, and dust storms can also result, as can an unmeasurable spread of human suffering.

When lands are best suited for a valuable export crop, ideally they might be planted to it and the proceeds then be channeled back into economic activities that provide jobs and increasing income to the rural poor. Instead, the peasants are all too frequently left with the worst of all worlds. The more productive lands are pre-empted by a few local or foreign investors and diverted to crops meant for sale abroad. The profits from this commercial agriculture wind up financing the luxurious lifestyle of the local landed gentry, being remitted abroad, or, if effectively taxed by the national government, mainly being spent in the cities to support bloated bureaucracies, urban services, and industrial development. Destitute subsistence farmers subsequently redouble their efforts on the remaining land, and ecological deterioration soon accelerates, as does the flood of migrants to urban shantytowns or less compacted neighboring countries.

The examples from Ethiopia and El Salvador each show, in different ways, why land redistribution may be the prerequisite of ecological recovery in countries with highly inequitable land tenure. Confidence in the personal benefits from land-improvement investments greatly increases the likelihood of such investments. Where the tenants are insecure, or where laborers are indentured to till the fields of another, guardianship of the long-term quality of the land can scarcely be expected. In some situations, the sheer problem of space is exacerbated by oversized estates in the hands of a small minority. Perhaps the most rigidly institutionalized examples of the latter are Rhodesia and South Africa, where the small European minorities have reserved the best lands for their own use, crowding Africans onto an inadequate land area.[7]

At the same time, however, it is naïve to assume that the political act of distributing land and other social benefits guarantees ecologically sound development. As the escalating deterioration of the Bolivian highlands since the reforms of the early 1950s demonstrates, breaking up the *latifundia* can actually exacerbate deforestation and erosion. After the French Revolution of 1789, a similar acceleration of environmental destruction occurred. Peasants and wood merchants immediately set upon the newly accessible forest reserves of the deposed feudal barons, destroying an estimated 3.5 million hectares of forest in just four years.[8] This phenomenon leads inescapably to an appreciation of the importance of *information* on proper and improper land uses, and of *ideas* about humanity's relationship to the land and its future. Even where legal structures and demographic and geographical conditions exist that would permit efficient, sustainable land exploitation, an *ethic* of conservation is its ultimate guarantor.

The blending together of social reforms and an ethical ecological consciousness sets the Chinese Revolution apart from most other major social movements of this or any century. The stress on land conservation that has permeated daily discourse in China is by no means an inevitable outgrowth of the Chinese leadership's European Marxist background. If the historical pattern of large-scale devastation is in fact being reversed in China, as the government claims and as many visitors attest, this may constitute the Chinese leadership's greatest legacy to future generations both in and out of their country. Without the massive efforts at reforestation and soil and water conservation, as well as the stringent programs to slow population growth that have been initiated, a future of destitution and famine for the world's most populous country would have been certain.

The degree to which land ownership patterns influence ecological trends varies widely among countries; the threats posed by *latifundia* in El Salvador bear only limited resemblance to the crises of Niger, India, or Nepal. There are, however, several other political obstacles to effective natural resource protection that exist in every country. One chronic problem, with implications extending beyond environmental issues, is the difficulty decision-making institutions have in adjudicating the competing demands of the present and the future —the problem of priorities. In the face of famines, military threats, unemployment, political intrigues, and other such everyday events in the average poor country, few powerful officials feel they can afford the "luxury" of devoting attention and money to a topic as seemingly esoteric as ecology. Until the environment suddenly *is* the major national crisis, its deterioration may occur almost unnoticed, with the costs quite real but difficult to total up. Likewise, the benefits of programs like tree planting and sapling protection, or the sharing of soil-conserving techniques with farmers on tiny holdings, are not easily appreciated by economic planners. Both governments and the public want dramatic production gains now, not an investment of scarce funds to satisfy what may seem hypothetical future needs. Hence emphasis is placed on big dams and canals at the expense of drainage; on clearing new farmland at the expense of reforesting depleted hillsides; and on building fertilizer factories at the expense of encouraging the preservation of soil fertility and structure with organic wastes.

Land-tenure systems as well as political considerations can promote a dangerous emphasis on the short-term future, and the subsequent effects on environmental quality are not confined to the countries with *latifundia*. In 1951, a top-level research group in the United States wrote:

> Tenure problems are one of the major "stumbling blocks" to the adoption of conservation practices in the Corn Belt. . . . Many farms in the Corn Belt are owned by absentee landlords who have little personal contact with their tenants. These owners do not realize that conservation adjustments will improve farm income over a period of several years. Instead, they want a high return on their investment *now*. On many farms the tenant is also interested in short-run profits. He may have only a one-year lease with no assurance of renewal, or the leasing agreement may require him to shoulder a larger share of the conservation costs than he receives in benefits.[9]

The statement stands in 1975.

An extreme case where tenure arrangements have induced myopic land treatment is Iran, particularly before the country's land reform program was initiated in 1962. Fully 62 percent of the country's farmers then rented their cropland, and it was common practice for landlords to shift tenants about each year from one field to another.[10] Under these conditions, tenants could hardly be expected to invest their energies in land improvements that would pay dividends only over the longer term.

The conflict between present and future needs can also have international ramifications. For example, many soil conservation officials in the United States winced in early 1974 when national agricultural and economic planners, confronting dwindling grain-reserve stocks, soaring prices, and a strong export market, called for "fence-to-fence planting." More than one old hand who had spent long, patient years coaxing Great Plains farmers to stop the cultivation of marginal areas found farmers again plowing up pasture lands in response to record wheat prices and the national leadership's appeal. Of course, economists charged with fighting inflation, as well as humanitarians supporting an expanded international food-relief effort, were happy to see every possible bushel of wheat produced. Yet, should wheat remain scarce for a prolonged period and overcultivation of the Great Plains again become chronic, the next extended drought will vividly document the self-defeating nature of this "solution" to grain scarcity.

These "priority" obstacles to sound land-use policies are in part problems of analysis and communications. Those best suited to do so have seldom tried to translate perceived ecological trends into the real costs accruing to society. While every government is busily adding up the hectares of new farmlands it has helped to bring into production, almost no one in any country is calculating the annual area *lost* to production due to erosion, salinization, or urbanization—and even fewer are totaling the ultimate social costs in lost output and compensatory new investments required to offset eroded or salted lands. The quantification of ecological losses and potential gains from recovery efforts is usually extremely difficult, but creative thinking in this area will be essential if decision-makers are to be convinced of the urgent need for shifts in priorities.

Even more crucial than the analytical failure has been the communications failure. Those who know the ecological score too frequently feel their job is done when a report is filed in a professional journal that is accessible, in terms of language as well as distribution, mainly to other scholars. Those who most need to know about impending ecological disasters and then act on their knowledge—particularly politicians, civil servants, journalists, and farmers—are frequently almost totally ignorant of what is happening. Establishing the urgency of a critical problem not traditionally recognized as such requires the constant broadcast of the facts and their implications to as many people as possible, through all possible means.

More effective analysis and communication of ecological problems may be a prerequisite of countering one of the greatest obstacles to a change in governmental priorities—the short time frame in which political leaders tend to operate. Ecologically sound planning requires concern for the next decade, the next generation and beyond; only the strong and vocal support—or insistence—of an informed citizenry can allow—or force—leaders to depart from their usual fixation on the next month or year. A widespread public understanding of the ecological danger is ultimately the prime weapon for fighting any commercial interests—whether highly placed timber concessionaires in Indonesia or Pakistan or corporate farmers in Central America—threatened by environmental protection measures. If powerful economic and political interests oppose necessary reforms, then a stronger political force is necessary to override them, and information about the nature of the threats to well-being is essential for building such a coalition. This is broadly true of virtually all political systems—not just democracies.

If saving the land pits the present against the future, in many ways it also pits the individual against society. Soil washed from the fields is a loss to the individual farmer but may cause even greater losses downstream as canals or reservoirs are choked with sediment. Yet those responsible for the damage are not required to pay the costs. The acquisition of more cattle on desert fringes may increase the wealth of the individual family or clan, and may even increase their personal chances of surviving a drought. But it may make ultimate disaster more likely for everyone. Moreover, any unnecessary harm to the quality of the land undermines the national heritage, whether or not the individual or corporate user views the land in a long-term perspective. In 1974, a group of leading American scientists described one way in which the clash between private and social interests is manifested in the United States: "The pressure of financial obligations encourages farming practices that result in excessive erosion. Farmers usually do not experience noticeable yield decreases because they have increased the use of technology which more than offsets the loss of production from excessive erosion. Land owners and farm operators thus tend to ignore long-term disinvestment damage to their land as well as off-site damages which may not affect them directly but which result in costs to the public."[11]

Once the changes in individual behavior necessary for the general welfare are identified—whether a reduction in herd sizes, a limit on fish catches, or the construction of terraces—the next hurdle is to secure compliance. While every situation has its unique complexities, a few general observations on this, perhaps the most intractable of all challenges to the environment, are worth repeating.

Experience has proven that sound treatment of the land cannot be decreed by officials—particularly those viewed as alien or oppressive—and then forced upon people who do not understand why changes in their habits are necessary. Faced with serious soil erosion in their African colonies, the British in the 1940s and 1950s tried the coercive approach, and by any account the ultimate results were abysmal. In parts of East Africa, eastern Nigeria, and elsewhere, colonial officials legislated changes in farming techniques and prohibited cultivation altogether in some watershed areas. Wise as the required measures may have been, the general response on the part of Africans was either resentment or apathy. A 1943 district commissioner's report in Kenya suggests the curious and untenable position in which the alien rulers found themselves: ". . . most of the people have no apparent intention of saving themselves and their descendants, and are indeed continually breaking new steeply sloped land as soon as one's back is turned."[12] Soil damage was temporarily arrested where the British overseers could directly impose their will, but soil conservation became identified with oppression, and, as African countries gained independence in the early 1960s, many of the anti-erosion programs were dropped.

In the United States, an elaborate voluntary program was established when the need for special attention to soil conservation was recognized. Thirteen thousand soil conservation officials in every corner of the nation offer advice to farmers on conservation needs and techniques and, in some areas, provide financial assistance for land improvements. While erosion remains an acute national problem, few dispute that the Soil Conservation Service has helped slow the accelerating land deterioration of the early part of this century —and without coercive measures. But if a voluntary program can have moderate success in the world's richest nation, where massive efforts can be financed, farmers are highly educated, and the cultural gaps between officials and farmers are generally slight, what conclusions might one draw for the poorest countries, where just the opposite conditions prevail? The dilemma is that where voluntary programs are least likely to be adequate, legislative fiats are least likely to change individual behavior. Throughout the poor countries there is already a gaping abyss between conservation laws and their enforcement. This reflects not only poor communications between governments and the rural poor, but also the clash between the exigencies of conservation and the individual's pressing, undeniable needs.

People hungry for land are not apt to leave forest or pasturelands unplowed, regardless of what ecological soundness dictates. Farmers hungry for bread are not likely to defer production this year to enhance soil quality for the next generation. Those with no other means than wood to cook their dinner cannot be expected to leave nearby trees unmolested even if they are labeled "reserved" by the government. And people brutalized by exploitive economic and social systems will probably not treat the land any more gently and respectfully than they are treated themselves.

Unfortunately, there are no quick solutions to the dismal cycles of poverty, ecological decay, and rapid population growth. To be sure, conservation ethics and problems need to be treated daily as "news" by the media and as part of basic curriculum in educational systems. Regulations protecting essential forests and mountain slopes also need to be strictly enforced. But these measures will never succeed until the populace has the technical and financial means to cooperate, and this means reaching the masses with ecologically sound agricultural advice and with credit facilities; maximizing rural employment on farms and in small-scale industries; and breaking down the social, legal, and economic structures that deny the poor basic opportunities for advancement. It means creating participatory institutions, whether through local government, cooperatives, or communes, that give the poor a sense of responsibility for and control over their destiny. That these prerequisites of ecological recovery are identical to the tactics of a more general war against poverty and hunger should come as no surprise.

Among a large share of the world's poor, words like "conservation" and "environmental protection," if they are known at all, strike a negative note. They are associated with denial and repression rather than with the improved quality of life that those who use them have in mind. Clearly, the movement to save an habitable environment will never succeed if its historical emphasis on protection and preservation is not balanced by progress in production and the satisfaction of basic human needs. Forestry departments will never effectively police forest reserves if they do not more successfully increase tree planting and wood production for local uses in forests, plantations, and the countryside. Soil conservation agencies will not stop the spread of cultivation to steep slopes if agricultural policies do not also increase production and employment on the better farmlands and improve the distribution of land ownership and production gains. The wildlife refuges and undisturbed natural ecosystems necessary in the interests of biological diversity, scientific study, and esthetics will never last if social conditions and productivity in adjacent areas are desperately low.

Conservation means protecting trees from the ax where necessary or desirable, but it also means far more—for the principal aspiration of the world's poorest half is to climb from the depths of severe social deprivation, not to save the environment for its own sake. Those concerned with global ecological deterioration and its consequences have no choice but to throw themselves into the maelstrom that is the politics of social change.

11. *Environmental Stress, Food, and the Human Prospect*

I̵ᴛ ᴡᴏᴜʟᴅ ʙᴇ a great mistake to view the various ecological trends discussed in this book as isolated, localized threats. Local threats they are, but unfortunately they also form a mosaic whose patterns help define many of the key global concerns of our age—issues which, directly or indirectly, will touch upon the lives of nearly everyone.

Few realize the extent to which the countries of the world are growing dangerously dependent on North America for food supplies, a trend to which the ecological deterioration of food-production systems contributes. As recently as the mid-1930s, Western Europe was the only continent with a grain deficit. Africa, Asia, Latin America, North America, the Soviet Union and Eastern Europe, and Australia all produced a small food surplus. To be sure, malnutrition was rampant among the poor of Africa, Asia, and Latin America, but in the commercial market each of these continents produced more food than the numbers and purchasing power of its inhabitants could consume.

The ensuing four decades have been marked by both general progress in reducing malnutrition in the world and a sorry attrition of sellers in the food market. One by one, continents have dropped from self-sufficiency to become net importers of food. By the mid-seventies, North America and Australia are the only regions with a net grain surplus, and North America controls a larger share of world grain exports than the Middle East does of world oil exports.

Europe, Japan, and the Soviet Union, all relatively affluent areas raising their consumption of grain-fed livestock products, are the major importers, but many of the poor countries, too, have gradually become more dependent on outside purchases to meet their minimum food needs as local production fails to keep pace with human numbers. A new current in the food market is the escalation of food imports by several of the oil-exporting countries, where, among at least part of the population, soaring incomes are boosting food demands far more rapidly than archaic and sometimes deteriorating agricultural systems can be modernized.

An extrapolation of current trends into the next decade and beyond leads in directions that bode ill for all countries, rich and poor. In their preparations for the 1974 World Food Conference, FAO analysts concluded that, if the production trends of the past decade continue, the expected growth of populations and incomes in the developing countries will produce a widening imbalance between food demand and production. Among the non-Communist developing countries, in fact, the need for grain imports may multiply fivefold between 1970 and 1985, reaching a net total of some eighty-five million tons per year.[1]

To the extent that the poor countries can fill this massive gap through food purchases abroad, claims on the exportable supplies of the few surplus countries—and especially the United States, the dominant global seller—will multiply. But as the world has learned since 1972, the chronic surplus-production capacity, large reserve stocks, and low, stable prices of the preceding two decades can no longer be taken for granted. Thus a growing drain on exportable supplies could well intensify inflationary pressures in all countries, as international demand pulls food prices up and forces costly investments that bring diminishing production returns in the agricultural sector of the advanced countries. The point could be reached where the sum of national grain import needs chronically exceeds the level North America is willing or able to supply, leaving heavy importers in a dangerous position. Furthermore, for the poor countries, the wholesale diversion of scarce foreign exchange from productive domestic investment to the purchase of food abroad would cripple economic development efforts.

Unpalatable as the latter prospect is, the more likely alternative is even more menacing. Given the past economic record and foreseeable economic future of many of the poorest countries, a good share of the potential food gap will probably be left unfilled by the commercial market. If so, scarcity will manifest itself in a rising incidence of malnutrition and premature death, the common assumption of steady historical progress toward a better life for all shattered. Under these circumstances the more affluent countries, particularly those with a food surplus, will face choices and responsibilities so politically sensitive that they may not be able to deal with them rationally. What portion of the exportable food should be reserved for charity, what portion for cash customers? Should domestic consumers alter their diets to make food available for the impoverished abroad? Are food gifts to needy countries moral or even responsible if they encourage greater tragedy in the future? Bitter debates over questions like these have already broken out in the United States; should international scarcity persist or recur and should food aid needs multiply over present levels, these thorny issues will become all the more acute and divisive—and the lives of more and more human beings will hang on the answers given.

Hearing talk of food shortages and future production constraints, some scientists calculate reassuringly that, with present-day technology put to work on all potentially arable lands, planet earth could feed fifteen, twenty, or even forty billion inhabitants. But rarely does the real world of human events intrude upon theoretical computations wearing such a gaunt face as it does in the case of food.

As the cost of further yield gains in the more advanced countries becomes harder to bear, and as the fossil fuels that undergird the most productive known agricultural systems become dearer, visions of adequately feeding a world population doubling over the next forty years (let alone one redoubling after that) will likely grow dimmer. Even as capital and energy considerations hamper the realization of hypothetic agricultural potential, every ton of fertile topsoil unnecessarily washed away, every hectare claimed by desert sands, every reservoir filled with silt further drains world productivity and spells higher costs for future gains in output. The levels of social organization and technical sophistication required to extract higher yields from the land will also climb as the natural productivity of ecosystems is impaired.

For economic as well as political reasons, then, the loss or degradation of arable land anywhere concerns everyone. The ecological deterioration of agricultural systems is most severe in poor countries. But it must be viewed in the context of a world food economy from which the comfortable margin of surplus of previous decades seems to be disappearing; of a world in which inflation, caused in part by unstable food prices, has emerged as one of the key economic challenges of the decade; of a world in which extreme dependence on one geographic region for food exports dramatizes the fabled foolhardiness of placing all our eggs in one basket. In today's market conditions, a loss of productive capacity in Algeria, to take one example, has a direct effect on the world price of wheat. Decades of environmental degeneration, recently capped by drought, have forced Algeria to purchase some two million tons of wheat—well over half the country's grain supply—on world markets in 1975.

Losses of productive capacity due to environmental stress must also be considered in the context of the reckless, inadequately measured takeover of current and potential farmlands by urban sprawl and other competing uses, a myopic activity occurring in both rich and poor countries. In the United States, for example, at least 240,000 hectares of cropland, including some land of the highest quality, are annually engulfed by urban and transportation development, reservoirs, and flood control programs. In overcrowded Egypt, over twenty-six thousand hectares of farmland are taken over each year by cities, roads, factories, and military installations—rivaling the new lands annually reclaimed for agriculture.[2] Land losses to nonagricultural uses join the losses to environmental deterioration to reduce the ability of our planet to produce food.

Discernible trends of deterioration, viewed against the ineluctable curve of population growth, raise the additional possibility that catastrophic agricultural collapses over large areas, causing famines and requiring major international emergency-relief efforts, will occur with increasing frequency. This contingency exists irrespective of any possible global climatic changes, which would by themselves raise the incidence of crop failures.

As human-caused stress on an ecosystem builds, whether on the hillsides of Java and Pakistan or the rangelands of Botswana and Afghanistan, the capacity of the vegetation and land to withstand

climatic extremes intact is exhausted. What might have been a difficult period of low rainfall becomes a period of famine and abandonment of once productive fields to desert sands; what might have been
a serious flood becomes a calamitous one, washing away a year's
harvest and a layer of fertile topsoil that took many centuries to build.
In mid-1975, large-scale famine-relief efforts are underway in parts of
Somalia, Ethiopia, and Haiti; in each case the immediate cause of
famine is drought, but in each case the stricken regions were ecological disaster zones well before the drought set in.

The site and nature of the next ecological catastrophe is impossible to predict, and we can only speculate about the human costs.
Likewise, we can at best surmise the impact of future ecological
debacles on the survival of governments; certainly the overturning of
several African regimes in the early seventies was at least partly linked
to famine conditions and food-relief politics in areas of ecological
distress. What we can say with confidence is that, when the next
calamities do occur, governments and the media will label them acts
of God when in fact humans will have helped set the stage.

To the extent that population growth in the afflicted countries is
slowed, ecological pressures will ease and their fearsome consequences will be at least postponed. Only a combination of improved
land-use habits with a drastic slowdown, and eventual halt, in the
population growth of Africa, Asia, and Latin America can put off
nature's day of reckoning altogether. The accumulating evidence of
severe ecological deterioration underscores the urgency of attacking
the population problem from all directions at once: making family
planning services universally available; liberating women from traditional roles; meeting basic social needs such as rudimentary health
care, adequate nutrition, and literacy that are usually associated with
reduced fertility; and reorienting social and economic policies to
promote smaller families.

The development and dissemination of renewable, decentralized,
and low-cost energy sources for the half of mankind now burning
wood, crop residues, or dung for fuel figures centrally in the ameliora-
tion of global environmental stress. Current energy research and
investment patterns, in the poor as well as rich countries, betray a
heavy preoccupation with new fuels for industry and the machines
of the rich, while the pressing energy crisis of the masses in the poor

countries is given short shrift. Extensive technical and sociological research, as well as ample funding for implementation, is needed to disseminate small-scale fuel sources such as solar cookers, bio-gas plants that convert dung into fuel and fertilizer, and, most of all, tree plantations.

Tree planting, like family planning, is needed nearly everywhere. Tree plantations, supporting sustainable fuel-wood industries, can provide badly needed jobs as they help curb the depletion of both remaining forests and scattered trees throughout the countryside of the poor nations. They can also help halt the increasingly frequent diversion of precious dung from fields to fireplaces. Tree planting on eroded hillsides, on abandoned farmlands, along roads, and between agricultural fields is insurance against the erosion of topsoils by wind and water, the accelerated choking of reservoirs and canals with silt, and the rising incidence of severe floods.

However crucial, reforestation and soil conservation programs cannot succeed without a concomitant transformation of agricultural methods on the lands best suited to agriculture. Many of the negative trends discussed in this book involve the spread of cultivation to marginal lands where no type of farming is sustainable; only the rapid expansion of food output and employment elsewhere, together with curbs on population growth, can curtail the futile exploitation of substandard lands and the razing of strategic forests.

Good land, too, is often damaged because its carrying capacity, its ability to support humans and animals on a sustainable basis, is surpassed. Yet the concept of agricultural carrying capacity takes on meaning only in conjunction with a particular technology. Properly managed, many areas now threatened with desertification or accelerated erosion could produce far more grain or meat than they do. Some of the institutional and political prerequisites of the needed agricultural revolution were noted in Chapter 10; they include, in most countries, reforms in land tenure and in the distribution of credit and advice, as well as a steadfast governmental commitment to helping the broad masses of farmers to improve their methods. Farm technologies tailored to local ecological conditions are also, of course, essential; research on appropriate farming systems, where such systems are not already known, ranks as a high priority.

A new, broader approach to development planning is required of both international development assistance agencies and national governments. Based on economic analyses that isolate a few threads from the whole cloth of natural environment and human activity, foreign aid projects and indigenous development programs alike too often fail to discern and eradicate the ecological roots of impoverishment.

"Environmental impact" assessments of proposed projects are a new stock-in-trade among many governments and aid agencies in the mid-seventies. Such investigations certainly represent a measurable step forward from the days, not long gone, when dams or factories could be built with hardly a thought to the harmful side effects that would cast shadows on the planned benefits. What is now needed, however, is another giant step beyond such assessments to the incorporation of an ecological perspective into the development-planning process from its inception. A planning exercise with the natural environment's capacity to serve human needs as its reference point would, in many countries, generate a different mix of priorities and projects than those supported by present-day systems. The need is not just for an agency to predict detrimental environmental consequences of projects chosen on the basis of traditionally quantified financial variables, but is, rather, for one to identify the programs and strategies needed to enhance the environment's ability to support an improved life for people. This is, after all, the ostensible goal of development.

Though nothing can substitute for a governmental commitment to ecological analysis and regeneration, and to agrarian reforms, some kinds of international assistance to the poorest countries are essential. The United States was able to recover the productivity of the Dust Bowl, and the Soviet Union that of the Virgin Lands, only because each country had the technical and financial resources needed to identify the sources of stress and then to act on their findings. Furthermore, both countries had enough momentum in their food economies—in the form of surplus output elsewhere or the ability to purchase grain abroad—to permit the necessary lag of farming efforts in the afflicted zones without serious shortages or prolonged hardship.

Many of the countries whose environments are most seriously threatened today, by contrast, are short of the multitude of technical skills—including, at a minimum, engineering, hydrology, forestry, agronomy, range management, and ecology—that must simultaneously be brought to bear on a disintegrating food system. And it is unrealistic to expect those eking out a precarious existence on degenerating lands to give their plots up to trees, to leave land fallow, or to sacrifice their herds if the government is unable to provide alternate sources of food and income. Channeling a higher share of international food aid into "food for work" programs, in which food wages support the poor while they rebuild the environment that poverty forced them to destroy, would make constructive use of the food aid that will be necessary in any case in coming years.

The systematic identification and analysis of trends in ecological deterioration, as well as the marshaling of technical and financial resources to oppose them, are the formidable tasks confronting the recently established United Nations Environment Program. The problems of "land, water, and desertification" have been accorded top priority by UNEP, and, by means of its new Global Environmental Monitoring System, UNEP will promote information gathering in various regions on such natural resource conditions as soil quality, deforestation rates, and oceanic pollution trends. Though far from comprehensive, the GEMS program is a meaningful start toward filling some glaring and dangerous gaps in humankind's accumulated knowledge about its milieu. The multitude of research efforts being sponsored by UNESCO's Program on Man and the Biosphere will similarly increase our knowledge of global ecological trends.

Yet the predictable slowness with which precise figures on ecological deterioration in one country or another become available—if they ever do—is no excuse for continued procrastination by political and economic decision-makers. Waiting for the ponderous process of scientific data-collection to produce definitive results would, for some countries, amount to committing ecological suicide. In too many areas, the spreading denudation of hillsides and overgrazing of rangelands is apparent to even the untrained eye; no computer print-outs, only an appreciation of the price humans will pay for inaction, should be necessary to justify initiating emergency salvage operations.

Identifying the underlying causes of the various symptoms of ecosystem overstress will tax to the fullest the analytical skills of governments and development planners. Falling agricultural yields, disappointing returns from capital investments and chemical fertilizer applications, and occasional dramatic disruptions of production over large areas will be the more obvious manifestations of unchecked deterioration. Tracing the resulting effects on nutritional status, economic prospects, political stability, and, indeed, on the very social fabric of whole societies is rather more difficult. The causes of social maladies are often obfuscated by the pressing demands of the symptoms. The indirect international economic and political effects, ranging from inflation to possible military conflicts, will likewise be shrouded by the particular events that catalyze them.

Seldom is a direct connection drawn between rural degeneration and mushrooming urban shantytowns, where degraded human conditions are creating a social tinderbox and stand out as one of the major social blights of the closing century. Even more seldom is that critical connection acted upon. Providing a decent, sustainable habitat for more humans in the countryside is far more practical and socially desirable than shouldering the almost insuperable task of providing a healthy habitat for burgeoning numbers in the cities, where no chance exists for productive universal employment. The tide of refugees can only be slowed through rural ecological regeneration; when topsoil washes down the mountainsides, those whose livelihoods depend on it lose their foothold.

The trends charted in this book do not point toward a sudden, cataclysmic global famine. What appears most likely, if current patterns prevail, is chronic depression conditions for the share of humankind, perhaps a fourth, that might be termed economically and politically marginal. Marginal people on marginal lands will slowly sink into the slough of hopeless poverty. Some will continue to wrest from the earth what fruits they can, others will turn up in the dead-end urban slums of Africa, Asia, and Latin America. Whether the deterioration of their prospects will be a quiet one is quite another question.

Notes

1. The Undermining of Food-Production Systems

1. Prominent examples in the English language include: G. V. Jacks and R. O. Whyte, *The Rape of the Earth—A World Survey of Soil Erosion* (London: Faber, 1939); Hugh Hammond Bennett, *Soil Conservation* (New York: McGraw Hill, 1939); W. C. Lowdermilk, "Man-Made Deserts," *Pacific Affairs*, Vol. 8, No. 4 (1935); Russell Lord, *Behold Our Land* (Boston: Houghton Mifflin, 1938); Fairfield Osborn, *Our Plundered Planet* (Boston: Little, Brown, 1948); Paul B. Sears, *Deserts on the March* (Norman: University of Oklahoma Press, 1935); William Vogt, *Road to Survival* (New York: William Sloane, 1948).
2. "The Nurturing Forest," *Ceres*, Vol. 8, No. 2 (March–April, 1975), p. 4.

2. A History of Deforestation

1. Anders Rapp, *A Review of Desertization in Africa—Water, Vegetation, and Man*, SIES Report No. 1 (Stockholm: Secretariat for International Ecology, 1974), p. 8.
2. Marvin W. Mikesell, "The Deforestation of Mount Lebanon," *Geographical Review*, Vol. 59, No. 1 (January, 1969). The following account of Mount Lebanon's history is largely drawn from this excellent article.
3. Herbert G. May and Bruce M. Metzger, eds., *The Oxford Annotated Bible* (New York: Oxford University Press, 1962), p. 421.
4. Omer C. Stewart, "Fire as the First Great Force Employed by Man," in William L. Thomas, Jr., ed., *Man's Role in Changing the Face of the Earth*, Vol. I (Chicago: University of Chicago Press, 1956), p. 118.
5. F. Fraser Darling, "Man's Ecological Dominance Through Domesticated Animals on Wild Lands," in William L. Thomas, ed., *op. cit.*, Vol. II, p. 779; Erhard Rostlund, "The Outlook for the World's Forests and Their Chief Products," in Stephen Haden-Guest, John K. Wright, and Eileen M. Teclaff, eds., *A World Geography of Forest Resources* (New York: Ronald Press, for the American Geographical Society, 1956), p. 640; and Hilgard O'Reilly Sternberg, "Man and Environmental Change in South America," in E. J. Fittkau, *et al.*, eds., *Biogeography and Ecology in South America*, Vol. 18, *Monographiae Biologicae*, ed. P. Van Oye (The Hague: Dr. W. Junk, 1968), p. 418.
6. Cited in Klaus Stern and Laurence Roche, *Genetics of Forest Ecosystems* (New York: Springer-Verlag, 1974), p. 240.

7. S. D. Richardson, *Forestry in Communist China* (Baltimore: Johns Hopkins University Press, 1966), p. 58.
8. D. Y. Lin, "China," in Stephen Haden-Guest, John K. Wright, and Eileen M. Teclaff, eds., *op. cit.*, p. 530.
9. H. C. Darby, "The Clearing of the Woodland in Europe," in William L. Thomas, ed., *op. cit.*, Vol. I.
10. Klaus Geissner, "Der Mediterrane Wald im Mahgreb," *Geographische Rundschau*, Vol. 23, No. 10 (October, 1971); and Klaus Müller-Hohenstein, "Die Anthropogene Beeinflussing der Walder im Westlichen Mittelmeerraum unter besonderer Berücksichtigung der Aufforstung," *Erdkunde: Archiv für Wissenshaftliche Geographie*, Vol. 27, No. 1 (March, 1973).
11. H. C. Darby, *op. cit.*
12. Peter Sartorius and Hans Henle, *Forestry and Economic Development* (New York: Praeger, 1968), p. 5.
13. Lesley Bird Simpson, *Many Mexicos* (Berkeley: University of California Press, 1964), p. 18.
14. Paul B. Sears, "The Importance of Forests to Man," in Stephen Haden-Guest, John K. Wright, and Eileen M. Teclaff, eds., *op. cit.*, p. 11; Paul B. Sears, quoted in William L. Thomas, ed., *op. cit.*, Vol. I, p. 406; and Tom Gill, *Land Hunger in Mexico* (Washington, D.C.: Charles Lathrop Pack Foundation, 1951), pp. 34–35.
15. W. C. Lowdermilk, "Man-Made Deserts," *Pacific Affairs*, Vol. 8, No. 4 (1935), p. 417.
16. Quoted in Peter Sartorius and Hans Henle, *op. cit.*, pp. 7–8.
17. Clarence J. Glacken, "Changing Ideas of the Habitable World," in William L. Thomas, ed., *op. cit.*, Vol. I, p. 77.
18. H. C. Darby, *op. cit.*, p. 204.
19. Erhard Rostlund, *op. cit.*, p. 639; and Food and Agriculture Organization, *World Forest Inventory, 1963* (Rome: 1965).
20. S. Kolade Adeyoju, "Forest Resources of Nigeria," *Commonwealth Forestry Review*, Vol. 53, No. 2, p. 102.
21. Food and Agriculture Organization, *Wood: World Trends and Prospects*. Basic Study No. 16 (Rome: 1967), p. 54.
22. Otto Soemarwoto, "The Soil Erosion Problem in Java," presented to First International Congress of Ecology, The Hague, September, 1974 (Bandung: Institute of Ecology, Padjadjaran University), 1974.
23. Pieter Lieftinck, A. Robert Sadove, and Thomas C. Creyke, *Water and Power Resources of West Pakistan: A Study in Sector Planning*, Vol. II (Baltimore: Johns Hopkins University Press, for the World Bank, 1969), p. 93.
24. Government of India, Ministry of Irrigation and Power, *Fifth Plan Position Paper on Flood Control, Drainage and Anti-Waterlogging, Anti-Sea Erosion* (New Delhi: May, 1972); Government of India, Ministry of Irrigation and Power, *Report of the Irrigation Commission, 1972*, Vols. I and II (New Delhi: 1972); Herbert Francis Mooney, "The Problem of Shifting Cultivation with Special Reference to Eastern India, the Middle East, and Ethiopia," *Fifth World Forestry Congress Proceedings*, Vol. III (Seattle: 1960), p. 2021; Gordon Conway and Jeff Romm, *Ecology and Resource Development in Southeast Asia*, Report to the Ford Foundation, Office for Southeast Asia (New York: Ford Foundation, August, 1973), pp. 7–8; World Food Programme, "Project Summary: Reforestation and Watershed Protection in the Upper Solo River Basin," Indonesia, project No. 648, WFP/IGC: 18/7 Add. 7 (Rome: September 21, 1970); Adrian G. Marshall, "Conservation in West Malaysia: The Potential for International Cooperation," *Biological Conservation*, Vol. 5, No. 2 (April 1973), p. 136; J. C. Delwaulle, "Désertification de l'Afrique au Sud du Sahara," *Bois et Forêts des Tropiques*, No. 149 (May–June, 1973), p. 11; and Clarke Brooke, "Food Shortages in Tanzania," *Geographical Review*, Vol. 57, No. 3 (July, 1967), p. 347.

25. Food and Agriculture Organization, *Wood: World Trends and Prospects, op. cit.*, p. 53.
26. J. P. Lanly, "Régression de la Forêt Dense en Côte d'Ivoire," *Boits et Forêts des Tropiques*, No. 127 (September–October, 1969).
27. Paul W. Richards, "The Tropical Rain Forest," *Scientific American*, December, 1973, p. 67.
28. S. D. Richardson, *op. cit.*, p. 12.
29. S. D. Richardson, *op. cit.*, p. 63; and Peter Sartorius and Hans Henle, *op. cit.*, p. 86.
30. Peter Sartorius and Hans Henle, *op. cit.*, p. 87. See also Jack C. Westoby, "How the Chinese Learn About Forestry," *American Forests*, Vol. 81, No. 6 (June, 1975).

3. Two Costly Lessons: The Dust Bowl and the Virgin Lands

1. Thomas Frederick Saarinen, *Perception of the Drought Hazard on the Great Plains*, Department of Geography Research Paper No. 106 (University of Chicago, 1966), esp. pp. 14–18.
2. See Paul Sears, "A Empire of Dust," in John Harte and Robert H. Socolow, eds., *Patient Earth* (New York: Holt, Rinehart & Winston, 1971); and *The Future of the Great Plains*, Report of the Great Plains Committee (Washington, D.C.: Government Printing Office, December, 1936), pp. 3–5.
3. Vance Johnson, *Heaven's Tableland: The Dust Bowl Story* (New York: Farrar, Straus, 1947) vividly describes the dust storms of the 1930s. H. H. Finnell, *Depletion of High Plains Wheatlands*, U.S. Department of Agriculture Circular No. 871 (June, 1951), discusses the organic contents of the dust storms.
4. In Frank E. Smith, ed., *Conservation in the United States, A Documentary History: Land and Water, 1900–1970* (New York: Chelsea House, in association with Van Nostrand Reinhold, 1971), p. 438.
5. Paul Sears, *op. cit.*, p. 10.
6. W. C. Lowdermilk, "Man-Made Deserts," *Pacific Affairs*, Vol. 8, No. 4 (1935), p. 417.
7. *The Future of the Great Plains, op. cit.*, p. 1.
8. Kenneth E. Grant, "Erosion in 1973–74: The Record and the Challenge," *Journal of Soil and Water Conservation*, Vol. 30, No. 1 (January–February, 1975).
9. Council for Agricultural Science and Technology, *Conservation of the Land, and the Use of Waste Materials for Man's Benefits*. Prepared for Committee on Agriculture and Forestry, United States Senate, March 25, 1975. (Washington, D.C.: Government Printing Office, 1975), esp. pp. 16, 17. On windbreaks, also see General Accounting Office, *Action Needed to Discourage Removal of Trees That Shelter Cropland in the Great Plains*, RED–75–375 (Washington, D.C.: General Accounting Office, June 20, 1975).
10. The dimensions of the program are outlined in William A. Dando, *Grain or Dust: A Study of the Soviet New Lands Program, 1954–1963*, Ph. D. dissertation, University of Minnesota, Department of Geography, 1969; and Frank A. Durgin, Jr., "The Virgin Lands Programme, 1954–1960," *Soviet Studies*, Vol. 13, No. 3 (1961–62). I am grateful to John Tidd for his assistance with research and analysis of the Soviet Virgin Lands program.
11. W. A. Douglas Jackson, "The Virgin and Idle Lands Program Reappraised," *Annals of the Association of American Geographers*, Vol. 52, No. 1 (March, 1962), p. 76.
12. Basic issues of the debate over proper farming techniques in the Virgin Lands are presented in Werner G. Hahn, *The Politics of Soviet Agriculture, 1960–1970*

(Baltimore: Johns Hopkins University Press, 1972); and in an article by A. Barayev, *Izvestia*, May 31, 1963.
13. *Izvestia*, November 22, 1961. Translation by John Tidd.
14. Hahn, *op. cit.*, p. 112.
15. Details of erosion damages in S. S. Kabysh, "The Permanent Crisis in Soviet Agriculture," in Roy D. Laird, ed., *Soviet Agriculture: The Permanent Crisis* (New York: Frederick A. Praeger, 1965), pp. 166–168; and Hahn, *op. cit.*, p. 112.
16. See Hahn, *op. cit.*, for discussion of the political implications of the 1963 Virgin Lands failure. Khrushchev's Italian interview is cited in Naum Jasny, *Khrushchev's Crop Policy* (London: George Outram, 1964), p. 23.

4. Encroaching Deserts

1. U.S. Agency for International Development, Office of Science and Technology, *Desert Encroachment on Arable Lands: Significance, Causes, and Control*, TA/OST 72–10 (Washington, D.C.: August, 1972), p. 1.
2. M. Kassas, "Arid and Semi-Arid Lands: An Overview," in United Nations Environment Programme, *Overviews in the Priority Subject Area: Land, Water and Desertification*, UNEP/PROG/2 (Nairobi: February, 1975), pp. 1–2.
3. U. S. Agency for International Development, Office of Science and Technology, *op. cit.*, pp. 2, 3.
4. Samir I. Ghabbour, "Some Aspects of Conservation in the Sudan," *Biological Conservation*, Vol. 4, No. 3 (April, 1972), pp. 228–229.
5. M. Kassas, "Desertification versus Potential for Recovery in Circum-Saharan Territories," in Harold E. Dregne, ed., *Arid Lands in Transition* (Washington, D.C.: American Association for the Advancement of Science, 1970), p. 124.
6. Kai Curry-Lindahl, "Conservation Problems and Progress in Equatorial African Countries," *Environmental Conservation*, Vol. 1, No. 2 (Summer, 1974), pp. 119–121.
7. H. N. Le Houérou, "North Africa: Past, Present, Future," in Harold E. Dregne, ed., *op. cit.*, p. 240.
8. For two examples see P. H. T. Beckett and E. D. Gordon, "Land Use and Settlement Round Kerman in Southern Iran," *Geographical Journal*, Vol. 132, pt. 4 (December, 1966), p. 476–489; and Food and Agriculture Organization/U.N. Special Fund, *Survey of Land and Water Resources, Afghanistan, Vol. 1, General Report* (Rome: 1965), p. 22. Overgrazing throughout the Middle East is discussed in Marion Clawson, Hans H. Landsberg, and Lyle T. Alexander, *The Agricultural Potential of the Middle East* (New York: American Elsevier, 1971), esp. pp. 74–75, 127. The Sonora Desert is discussed in J. L. Cloudsley-Thompson, "Animal Utilization," in Harold E. Dregne, ed., *op. cit.*, p. 59.
9. Cesar F. Vergelin, "Water Erosion in the Carcarana Watershed: An Economic Study," Ph.D. dissertation, University of Wisconsin, 1971, p. 11; and U.S. Agency for International Development, Office of Science and Technology, *op. cit.*, p. 4.
10. Central Arid Zone Research Institute, "Sociology-Main Results," mimeographed (Jodhpur: n.d.), p. 2.
11. M. S. Swaminathan, *Our Agricultural Future*, Sardar Patel Memorial Lectures, October 30, 31, and November 1, 1973 (New Delhi: India International Centre, 1973), p. 22.
12. National Planning Commission, 1952, quoted in Government of India, National Commission on Agriculture, *Interim Report on Desert Development* (New Delhi: March, 1974), p. 3; B. B. Roy and S. Pandey, "Expansion or Contraction of the Great Indian Desert," in *Proceedings of the Indian National Science Academy*, Vol. 36, B, No. 6 (1970), p. 343. Desert expansion asserted in Reid A. Bryson and David A. Baerreis, "Possibilities of Major Climatic Modification and Their Im-

plications: Northwest India, A Case for Study," *Bulletin of the American Meteoro-logical Society*, Vol. 48, No. 3 (March, 1967), p. 141. The Indian government's 1974 *Interim Report on Desert Development*, cited above, also assumes a continu-ing spread of the desert.

13. Reid A. Bryson and David A. Baerreis, *op. cit.*

14. N. H. MacLeod, "Dust in the Sahel: Cause of Drought?" mimeographed (Wash-ington, D.C.: Drought Analysis Laboratory, American University, August, 1974); and Reid A. Bryson, *The Sahelian Effect*, University of Wisconsin, Institute for Environmental Studies, Working Paper No. 9 (Madison: August, 1973).

15. H. N. Le Houérou, "Deterioration of the Ecological Equilibrium in the Arid Zones of North Africa," mimeographed (Rome: Food and Agriculture Organiza-tion, 1974), p. 5; Anders Rapp, *A Review of Desertization in Africa—Water, Vegetation, and Man*, SIES Report No. 1 (Stockholm: Secretariat for Interna-tional Ecology, 1974), pp. 13–28; B. B. Roy and S. Pandey, *op. cit.;* and personal correspondence from Reid A. Bryson, June 27, 1975.

16. "Statement of the IFIAS Workshop on the Impact of Climatic Change on the Quality and Character of Human Life," adopted in Bonn, May 10, 1974 (Stock-holm: International Federation of Institutes for Advanced Study, June 3, 1974); Tom Alexander, "Ominous Changes in the World's Weather," *Fortune*, Febru-ary, 1974; and Reid Bryson, *op. cit.*

17. See F. Fraser Darling and Mary A. Farvar, "Ecological Consequences of Sedentar-ization of Nomads," in M. Taghi Farvar and John P. Milton, eds., *The Careless Technology: Ecology and International Development* (New York: Natural History Press, 1972); and Jeremy Swift, "Disaster and a Sahelian Nomad Economy," in David Dalby and R. J. Harrison Church, eds., *Drought in Africa, Report of the 1973 Symposium*, Centre for African Studies, University of London (London: 1973).

18. Lee M. Talbot, "Ecological Consequences of Rangeland Development in Masai-land, East Africa," in M. Taghi Farvar and John P. Milton, eds., *op. cit.*

19. U. S. Agency for International Development, "An Approach to the Recovery and Stabilization of the Sahelian-Sudanian Range and Livestock Industry," AID/AFR/CWR Technical Staff Paper, draft (Washington, D.C.: January, 1974).

20. N. H. MacLeod, *op. cit.*, p. 16.

5. Refugees from Shangri-La:
Deteriorating Mountain Environments

1. "Highlands" is defined as all areas over one thousand meters in altitude.

2. UNESCO, Programme on Man and Biosphere, *Working Group on Project 6: Impact of Human Activities on Mountain and Tundra Ecosystems, Final Report*, Lillehammer, November, 20–23, 1973, MAB Report Series No. 14 (Paris: 1974), pp. 20–21.

3. International Workshop on the Development of Mountain Environment, press release, Munich, December 12, 1974 (Feldafing: German Foundation for Interna-tional Development, December, 1974).

4. R. G. M. Willan, Chief Conservator of Forests, *Forestry in Nepal.* (Kathmandu: UNDP, 1967); and Government of Nepal, National Planning Commission Secre-tariat, *Draft Proposals of Task Force on Land Use and Erosion Control* (Kath-mandu: August, 1974).

5. International Bank for Reconstruction and Development/International Develop-ment Association (IBRD/IDA), *Economic Situation and Prospects of Nepal*, April 13, 1973, and *Sector Annexes*, August 15, 1973 (Washington, D.C.); and Tribhuvan University, Centre for Economic Development and Administration, *Regional Development Study (Nepal)* (Kirtipur, Kathmandu: May, 1973–Decem-ber, 1974), p. 3.

6. Government of India, Department of Agriculture, *Soil and Water Conservation in Nepal: Report of the Joint Indo-Nepal Team* (New Delhi: November, 1967), p. 5.

7. John C. Cool, *The Far Western Hills: Some Longer Term Considerations*, mimeographed (U.S. Agency for International Development, February, 1967), p. 6.

8. Government of Nepal, National Planning Commission Secretariat, *op. cit.*

9. *Ibid.*

10. IBRD/IDA, *Agricultural Sector Survey, Nepal, Vol. II* (Washington, D.C.: September 3, 1974), p. 7.

11. Government of Nepal, Ministry of Forests, "Introduction of Department of Soil and Water Conservation," mimeographed (Kathmandu, n.d.); Government of Nepal, National Planning Commission Secretariat, *op. cit.;* and personal interviews with Nepalese forestry officials, March, 1975.

12. Ernest Robbe, *Report to the Government of Nepal on Forestry*, ETAP Report No. 209 (Rome: Food and Agriculture Organization), March, 1954; and R. G. M. Willan, *op. cit.*, p. 16.

13. Government of Nepal, National Planning Commission Secretariat, *op. cit.*

14. Government of India, Department of Agriculture, *op. cit.*

15. D. C. Kaith, "Forest Practices in Control of Avalanches, Floods, and Soil Erosion in the Himalayan Front,". *Fifth World Forestry Congress Proceedings*, Vol. III (Seattle: 1960).

16. World Food Programme, *Interim Evaluation of Project Pakistan 385—"Watershed Management in the Kaghan and Daur Valleys,"* draft, October 29, 1974 (Rome: Food and Agriculture Organization, October, 1974).

17. FAO/U.N. Special Fund, *Survey of Land and Water Resources, Afghanistan, Vol. 1., General Report* (Rome: 1965), p. 22; and K. J. Lampe, *Forst in Paktia, Afghanistan* (Frankfurt/Mein: Federal Agency for Development Assistance), 1972.

18. William H. McNeill, *The Rise of the West* (New York: Mentor, 1963), p. 455.

19. J. Alden Mason, *The Ancient Civilizations of Peru* (Harmondsworth: Penguin, 1957), p. 137.

20. F. Monheim, "The Population and Economy of Tropical Mountain Regions, Illustrated by the Examples of the Bolivian and Peruvian Andes," presented to International Workshop on the Development of Mountain Environment, Munich, December 8–14, 1974 (Feldafing: German Foundation for International Development, December, 1974), p. 6; and J. Alden Mason, *op. cit.*, p. 139.

21. F. Monheim, *op. cit.*, p. 7; J. Alden Mason, *op. cit.*, p. 134. Alfred Metraux, *The History of the Incas*, translated by George Ordish (New York: Random House, 1969), p. 166.

22. Alfred Metraux, *op. cit.*, pp. 166–174, 187–196, discusses the exploitation of the Andean Indians; quote from J. Alden Mason, *op. cit.*, p. 135.

23. F. Monheim, *op. cit.*, p. 7; J. Alden Mason, *op. cit.*, p. 138; and R. F. Watters, *Shifting Cultivation in Latin America* (Rome: Food and Agriculture Organization, 1971).

24. R. F. Watters, *op. cit.*, pp. 35, 119, 135, 222, 225, 228.

25. David A. Preston, "The Revolutionary Landscape of Highland Bolivia," *Geographical Journal*, Vol. 135, pt. 1 (March, 1969), pp. 1–16.

26. R. F. Watters, *op. cit.*, pp. 230, 213.

27. F. Monheim, *op. cit.*, p. 8; and República del Perú, Oficina Nacional de Evaluación de Recursos Naturales, and Organización de los Estados Americanos. *Lineamientos de Política de Conservación de los Recursos Naturales Renovables del Perú* (Lima: May, 1974), p. 18.

28. James A. Liggett, "Erosion, Landslides, and Sedimentation in Colombia," proposal to the U.S. Department of State, Agency for International Development-Bogota, by Cornell University, Corporación Antónoma del Valle de Cauca, and Universidad del Valle, December, 1974.

29. Cited in R. F. Watters, *op. cit.*, p. 36.

194   (pages 91–97) Notes

30. Philip Ainsworth Means, *Ancient Civilizations of the Andes* (New York: Gordian Press, 1964; reprint of 1931 edition), p. 12; and República del Perú, Oficina Nacional de Evaluación de Recursos Naturales, and Organización de los Estados Americanos, *op. cit.*, p. 23.
31. F. Monheim, *op. cit.*, p. 10., provides migration data for Peru.
32. República del Perú, Oficina Nacional de Evaluación de Recursos Naturales, and Organización de los Estados Americanos, *op. cit.*, pp. 8–14; and R. F. Watters, *op. cit.*, p. 138.
33. República del Perú, Oficina Nacional de Evaluación de Recursos Naturales, and Organización de los Estados Americanos, *op. cit.*, pp. 8–14.
34. Gildas Nicolas, "Peasant Rebellions in the Socio-Political Context of Today's Ethiopia," presented at the Fifteenth Annual Meeting of the African Studies Association, Philadelphia, November 8–11, 1972, p. 11.
35. World Food Programme, "Project Summary: Reforestation and Soil Conservation in the Province of Wollo," Ethiopia Project No. 2097, WFP/IGC: 26/9 Add. 11 (Rome: July, 1974); World Food Programme, "Reforestation and Soil Conservation in the Province of Tigre," Ethiopia Project No. 769, WFP/IGC: 23/10 Add. 2 (Rome: Food and Agriculture Organization, March, 1973); U.S. Department of the Interior, Bureau of Reclamation, *Land and Water Resources of the Blue Nile Basin, Ethiopia, Appendix VI, Agriculture and Economics*, prepared for U.S. Agency for International Development (Washington, D.C.: 1964), p. 140; Herbert Francis Mooney, "The Problem of Shifting Cultivation with Special Reference to Eastern India, the Middle East, and Ethiopia," *Fifth World Forestry Congress Proceedings*, Vol. III (Seattle: 1960), p. 2023; quote from W. D. Ware-Austin, "Soil Erosion in Ethiopia: Its Extent, Main Causes and Recommended Remedial Measures," mimeographed (Addis Ababa: Institute of Agricultural Research, May, 1970), p. 1.
36. Leslie Brown, *East African Mountains and Lakes* (Nairobi: East African Publishing House, 1971), pp. 70–71.
37. Special Fund, *Report on Survey of the Awash River Basin, Vol. II, Soils and Agronomy* (Rome: 1965), p. 114.
38. *Ibid.*, p. 116; see also Leslie Brown, *op. cit.*, pp. 72, 73.
39. Richard St. Barbe Baker, "Some Reflections on Trees and Forests for Ethiopia," *Ethiopian Observer*, Vol. VIII, No. 2 (1964), p. 190; and Ronald J. Horvath, "Addis Ababa's Eucalyptus Forest," *Journal of Ethiopian Studies*, Vol. VI, No. 1 (January, 1968), pp. 13–19.
40. R. T. Jackson, P. M. Mulvaney, T. P. J. Russell, and J. A. Forster, *Report of the Oxford University Expedition to the Gamu Highlands of Southern Ethiopia, 1968* (Oxford: School of Geography, Oxford University, 1968), pp. 34–37.
41. See Gildas Nicolas, *op. cit.*; John M. Cohen, "Land Reform in Ethiopia: The Effects of an Uncommitted Center on the Rural Periphery," presented at the Sixteenth Annual Meeting of the African Studies Association, Syracuse, October 31–November 3, 1973; and William A. Hance, *The Geography of Modern Africa* (New York: Columbia University Press, 1964), pp. 351–364.
42. A. Warren, "East Africa," in B. W. Hodder and D. R. Harris, eds., *Africa in Transition* (London: Methuen, 1967), p. 117; M. S. Parry, "Progress in the Protection of Stream-Source Areas in Tanganyika," *East African Agricultural and Forestry Journal*, Vol. XXVII (special issue, March, 1962), p. 104; William A. Hance, *op. cit.*, pp. 378, 395–411; and W. J. Lusigi, "Some Environmental Factors in Food Production in Kenya," prepared for the United Nations World Food Conference, November 5–16, 1974 (Nairobi: National Environment Secretariat, Office of the President), 1974.
43. Paul H. Temple, "Soil and Water Conservation Policies in the Uluguru Mountains, Tanzania," and Anders Rapp, *et al.*, "Soil Erosion and Sediment Transport in the Morogoro River Catchment, Tanzania, both in Anders Rapp, Len Berry, and Paul Temple, eds., *Studies of Soil Erosion and Sedimentation in Tanzania*

(University of Dar es Salaam, Bureau of Resource Assessment and Land Use Planning and University of Uppsala, Department of Physical Geography, 1972), pp. 110–155.
44. *Ibid.*
45. World Food Programme, *Interim Evaluation of Project Pakistan 385—"Watershed Management in the Kaghan and Daur Valleys,"* op. cit.
46. República del Perú, Oficina Nacional de Evaluación de Recursos Naturales, and Organización de los Estados Americanos, op. cit., p. 20.

6. The Other Energy Crisis: Firewood

1. Keith Openshaw, "Wood Fuels the Developing World," *New Scientist*, Vol. 61, No. 883 (January 31, 1974). See also Food and Agriculture Organization, *Wood: World Trends and Prospects*, Basic Study No. 16 (Rome: 1967) for a brief overview of world fuel wood trends.
2. J. C. Delwaulle, "Désertification de l'Afrique au Sud du Sahara," *Bois et Forêts des Tropiques*, No. 149 (Mai–Juin, 1973), p. 14; and Victor D. DuBois, *The Drought in West Africa*, American Universities Field Staff, West African Series, Vol. XV, No. 1 (1974).
3. India's enforcement efforts are discussed in Government of India, Ministry of Agriculture, *Interim Report of the National Commission on Agriculture on Social Forestry* (New Delhi: August, 1973), p. 37. Tree-protection problems in China noted in S. D. Richardson, *Forestry in Communist China* (Baltimore: Johns Hopkins University Press, 1966), pp. 14, 66.
4. See, for example, Lila M. and Barry C. Bishop, "Karnali, Roadless World of Western Nepal," *National Geographic*, Vol. 140, No. 5 (November, 1971), p. 671.
5. Government of India, Ministry of Agriculture, op. cit.; S. K. Adeyoju and E. N. Enabor, *A Survey of Drought Affected Areas of Northern Nigeria* (University of Ibadan, Department of Forestry, November, 1973), p. 48; U.S. Department of the Interior, Bureau of Reclamation, *Land and Water Resources of the Blue Nile Basin, Ethiopia, Appendix VI, Agriculture and Economics* (Washington, D.C.: 1964), p. 12; Leslie H. Brown, *Conservation for Survival: Ethiopia's Choice* (Haile Selassie I University, 1973), pp. 63, 64; FAO/UNDP, *Forestry Research, Administration, and Training, Arbil, Iraq*, Technical Report I, "A Forest Improvement Programme," FAO/UNDP, IRQ-18 TR 1, IRQ/68/518 (Rome: Food and Agriculture Organization, 1973); República del Perú, Oficina Nacional de Evaluación de Recursos Naturales, and Organización de los Estados Americanos, *Lineamientos de Política de Conservación de los Recursos Naturales del Perú* (Lima: May, 1974), p. 18; and J. Alden Mason, *The Ancient Civilizations of Peru* (Harmondsworth: Penguin, 1957), p. 137.
6. FAO/UNDP, *Agricultural Survey and Demonstration in Selected Watersheds, Republic of Korea, Vol. I, General Report*, FAO/SF: 47/KOR 7 (Rome: Food and Agriculture Organization, 1969), p. 17; and Donald E. Vermeer, "Population Pressure and Crop Rotational Changes among the Tiv of Nigeria," *Annals of the Association of American Geographers*, Vol. 60, No. 2 (June, 1970), p. 311.
7. Government of India, Ministry of Agriculture, op. cit.
8. John Woodward Thomas, "Employment Creating Public Works Programs: Observations on Political and Social Dimensions," in Edgar O. Edwards, ed., *Employment in Developing Nations* (New York: Columbia University Press, 1974), p. 307.
9. See "Potentials for Solar Energy in the Sahel," interview with A. Moumoumi, *Interaction* (Washington, D.C.), Vol. III, No. 10, (July, 1975); National Academy of Sciences, Office of the Foreign Secretary, *Solar Energy in Developing Coun-*

tries: Perspectives and Prospects (Washington, D.C.: March, 1972); Farrington Daniels, *Direct Use of the Sun's Energy* (New York: Ballantine, 1974; reprint of 1964 edition); Denis Hayes, "Solar Power in the Middle East," *Science*, Vol. 188, No. 4195 (June 27, 1975), p. 1261.

10. C. R. Prasad, K. Krishna Prasad, and A. K. N. Reddy, "Bio-Gas Plants: Prospects, Problems and Tasks," *Economic and Political Weekly* (New Delhi), Vol. IX, Nos. 32–34 (special issue, August, 1974); and Arjun Makhijani with Alan Poole, *Energy and Agriculture in the Third World.* (Cambridge, Mass.: Ballinger, 1975), esp. ch. 4.

11. E. F. Schumacher, "Buddhist Economics," in *Small Is Beautiful: Economics as If People Mattered* (New York: Harper & Row, 1973).

7. The Salting and Silting of Irrigation Systems

1. W. C. Lowdermilk, "Conquest of the Land Through Seven Thousand Years," U.S. Department of Agriculture, Agriculture Information Bulletin No. 99, August, 1953.

2. Much of this section on Mesopotamia is based on Thorkild Jacobsen and Robert M. Adams, "Salt and Silt in Ancient Mesopotamian Agriculture," *Science*, Vol. 128, No. 3334 (November 21, 1958). See also William H. McNeill, *The Rise of the West* (New York: Mentor, 1963); Tom Dale and Vernon Gill Carter, *Topsoil and Civilization* (Norman: University of Oklahoma Press, 1955); R. van Aart, "Drainage and Land Reclamation in the Lower Mesopotamian Plain," *Nature and Resources*, Vol. X, No. 2 (April–June, 1974); P.J. Dielman, ed., *Reclamation of Salt Affected Soils in Iraq* (Wageningen, Netherlands: International Institute for Land Reclamation and Improvement, Publication 11, 1963); Marion Clawson, Hans H. Landsburg, and Lyle T. Alexander, *The Agricultural Potential of the Middle East* (New York: American Elsevier, 1971); and Food and Agriculture Organization, *Salinity Seminar, Baghdad* (Rome: Irrigation and Drainage Paper 7, 1971).

3. White House—Department of the Interior Panel on Waterlogging and Salinity in West Pakistan, *Report on Land and Water Development in the Indus Plain* (Washington, D.C.: Government Printing Office, 1964). See also D. W. Greenman, W. V. Swarrenski, and G. D. Bennett, *Ground-Water Hydrology of the Punjab, West Pakistan with Emphasis on Problems Caused by Canal Irrigation*, U.S. Geological Survey Water Supply Paper 1608–H (Washington, D.C.: Government Printing Office, 1967); Aloys Arthur Michel, *The Indus Rivers: A Study of the Effects of Partition* (New Haven: Yale University Press, 1967), esp. pp. 455–514.

4. Water and Power Development Authority, West Pakistan, *Programme for Waterlogging and Salinity Control in the Irrigated Areas of West Pakistan* (Lahore: May, 1961).

5. U.S. Agency for International Development, "Progress and Evaluation, SCARP-1, West Pakistan," April, 1965.

6. Water and Power Development Authority, Pakistan, "Accelerated Programme of Waterlogging and Salinity Control in Pakistan," draft (Lahore: February, 1975); and Water Resources Sector, Planning Commission, Government of Pakistan "Preliminary Proposals for an Accelerated Programme of Waterlogging and Salinity Control in West Pakistan" (Islamabad: 1973).

7. Roman W. Szechowycz and M. Mohsin Qureshi, "Sedimentation in Mangla Reservoir," *Journal of the Hydraulics Division* (Proceedings of the American Society of Civil Engineers), September, 1973.

8. See Pietier Lieftinck, A. Robert Sadove, and Thomas C. Creyke, *Water and Power Resources of West Pakistan: A Study in Sector Planning*, Vols. 1 and 2 (Baltimore:

Johns Hopkins Press, for the World Bank, 1968); and Aloys Arthur Michel, *The Indus Rivers: A Study of the Effects of Partition, op. cit.*

9. Aloys A. Michel, "The Impact of Modern Irrigation Technology in the Indus and Helmand Basins of Southwest Asia," in M. Taghi Farvar and John P. Milton, eds., *The Careless Technology: Ecology and International Development* (Garden City, N.Y.: Natural History Press, 1972), discusses some of the factors contributing to an anti-drainage bias.

10. Ronald C. Reeve and Milton Fineman, "Salt Problems in Relation to Irrigation", in Robert M. Hagen, Howard R. Haise, and Talcott W. Edminister, eds., *Irrigation of Agricultural Lands* (Madison, Wis.: American Society of Agronomy, Agronomy Series No. 11, 1967), p. 988; V. Kovda, "The Management of Soil Fertility," *Nature and Resources*, Vol. VIII, No. 2 (April–June, 1972), p. 3; and V. Kovda, "The World's Soils and Human Activity," in Nicholas Polunin, ed., *The Environmental Future* (New York: MacMillan, 1972), p. 374.

11. Government of India, Ministry of Irrigation and Power, *Report of the Irrigation Commission*, Vol. I (New Delhi: 1972), pp. 308–315; B. B. Vohra, *A Charter for the Land* (New Delhi: Ministry of Agriculture, September, 1972); Sudhir Sen, *A Richer Harvest: New Horizons for Developing Countries* (Maryknoll, New York: Orbis, 1974), pp. 214–215, 232; and Leslie T. C. Kuo, *The Technical Transformation of Agriculture in Communist China* (New York: Praeger, 1972), pp. 123–125.

12. Food and Agriculture Organization, *Salinity Seminar, Baghdad, op. cit.*, pp. 26, 53, 215.

13. *Ibid.*, pp. 200, 203.

14. Hilgard O'Reilly Sternberg, "Man and Environmental Change in South America," in E. J. Fittkau, *et al.*, eds. *Biogeography and Ecology in South America*, Vol. 18, *Monographiae Biologicae*, ed. P. Van Oye (The Hague: Dr. W. Jungk, 1968), p. 432; República del Perú, Oficina Nacionál de Evaluación de Recursos Naturales, and Organización de los Estados Americanos, *Lineamientos de Política de Conservacion de los Recursos Naturales del Perú* (Lima: May, 1974), p. 22; Arturo Cornejo T., "Resources of Arid South America," in Harold E. Dregne, ed., *Arid Lands in Transition* (Washington, D.C.: American Association for the Advancement of Science, 1970), p. 370; IBRD/IDA, *Agricultural Sector Study, Vol. IV, Agricultural Development in Brazil*, PA 52a (Washington, D.C.: August 19, 1970), p. 249; and FAO/UNDP, *Haiti: Rapport Final, Vol. IV, Genie Rural*, FAO/SF:45/HAI-3 (Rome: 1969), pp. 13–19.

15. David Henderson, "Arid Lands Under Agrarian Reform in Northwest Mexico," *Economic Geography*, Vol. 41, No. 4 (October, 1965), pp. 304–306; and Efraim Hernández Xolocotzi, "Mexican Experience," in Harold E. Dregne, ed., *op. cit.*, p. 324.

16. *International Symposium on the Salinity of the Colorado River*, an entire issue of *Natural Resources Journal* (University of New Mexico School of Law), Vol. 15, No. 1 (January, 1975), contains useful accounts of the river salinity problems and United States-Mexican diplomacy on the subject.

17. Myron B. Holburt and Vernon E. Valantine, "Present and Future Salinity of Colorado River," *Journal of the Hydraulics Division* (Proceedings of the American Society of Civil Engineers), March, 1972; and U.S. Department of the Interior, *Report on Water for Energy in the Upper Colorado Basin* (Washington, D.C.: July, 1974).

18. Harold E. Dregne, "Salinity Aspects of the Colorado River Agreement," in *International Symposium on the Salinity of the Colorado River, op. cit.*, p. 51.

19. S. J. Ursic and Farris E. Dendy, "Sediment Yields from Small Watersheds Under Various Land Uses and Forest Covers," in *Proceedings of the Federal Inter-Agency Sedimentation Conference, 1963*, U.S. Department of Agriculture, Agricultural Research Service, Miscellaneous Publication No. 970 (Washington, D.C.: June, 1965), p. 48.

20. Robert N. Allen, "The Anchicaya Hydroelectric Project in Colombia: Design and

Sedimentation Problems," in M. Taghi Farvar and John P. Milton, eds., *op. cit.*

21. "Shihmen Reservoir Catchment Area Management," in Economic Commission for Asia and the Far East, *Proceedings of the Ninth Session of the Regional Conference on Water Resources Development in Asia and the Far East*, Water Resources Series No. 40 (New York: United Nations, 1971); and Juan L. Mercado, "Formula for National Suicide," *Honolulu Star-Bulletin*, April 22, 1971, quoting the Philippine forester Nicolas P. Lansigan.

22. "Satellite Data, A Resource Management Tool for the Niger Watershed. Sub-Project 1," proposal by the Resource Analysis Laboratory, Biology Department, American University, February 6, 1975.

23. B. B. Vohra, *A Charter for the Land, op. cit.*, p. 7; and Government of India, Ministry of Irrigation and Power, *Report of the Irrigation Commission, 1972* (New Delhi: 1972), esp. ch. XIV, "Sedimentation of Reservoirs."

24. William G. Hoyt and Walter B. Langbein, *Floods* (Princeton: Princeton University Press, 1955), pp. 154–155; and John B. Stoll, "Soil Conservation Can Reduce Reservoir Sedimentation," *Public Works*, Vol. 93, No. 9 (September, 1962), p. 125.

25. Sumitro Djojohadikusumo, *Indonesia Towards the Year 2000* (Jakarta: Ministry of Research, February, 1975). Figures for Java include the neighboring Island of Madura. See also Denys C. Schwaar and John F. Harrop, "Agricultural Development and Population Growth in Indonesia," draft (Food and Agriculture Organization, Indonesia, 1974).

26. Otto Soemarwoto, "The Soil Erosion Problem in Java," presented to First International Congress of Ecology, The Hague, September 1974 (Bandung: Institute of Ecology, Padjadjaran University, 1974); Jac. P. Thijsse, "Will Java Become a Desert?" mimeograph, 1974; International Bank for Reconstruction and Development/International Development Association, *Development Issues for Indonesia*, Vol. II, Annex 3, "The Agricultural Sector" (Washington, D.C.: December 1, 1972), p. 5.

I am grateful to Boyd Compton for his field research and insights on the ecological problems of Southeast Asia. Uncited estimates of forest cover in Indonesia, the Philippines, and Thailand are based on his confidential interviews and personal examinations of satellite imagery.

27. Lester R. Brown with Erik P. Eckholm, *By Bread Alone* (New York: Praeger, 1974), esp. ch. 7.

8. *Myth and Reality in the Humid Tropics*

1. A. C. S. Wright and J. Bennema, *The Soil Resources of Latin America*, FAO/UNESCO Project, World Soil Resources Report No. 18 (Rome: Food and Agriculture Organization, 1965), p. 113.

2. An outstanding review of the biological aspects of shifting cultivation is P. H. Nye and D. J. Greenland, *The Soil Under Shifting Cultivation*, Technical Communication No. 51, Farnham Royal (Bucks: Commonwealth Agricultural Bureaux, 1960).

3. Quoted in Hans Ruthenberg, *Farming Systems in the Tropics* (Oxford: Clarendon Press, 1971), p. 53.

4. William Allan, *The African Husbandman* (New York: Barnes & Noble, 1965), p. 441. See also R. de Coene, "Agricultural Settlement Schemes in the Belgian Congo," *Tropical Agriculture*, Vol. 33, No. 1 (January, 1956), for a review of the corridor system.

5. William Allan, *op. cit.*, ch. 21; A. T. Grove, "Population Densities and Agriculture in Northern Nigeria," in K. M. Barbour and R. M. Prothero, eds., *Essays on African Population* (New York: Praeger, 1962); John M. Hunter, "Population

Pressure in a Part of the West African Savanna: A Study of Nangodi, Northeast Ghana," *Annals of the Association of American Geographers*, Vol. 57, No. 1 (March, 1967); N. H. MacLeod, "Use of ERTS Imagery and other Space Data for Rehabilitation and Development Programs in West Africa," in *Earth Resources Survey System*, Hearings Before the Subcommittee on Space Science and Applications of the Committee on Science and Astronautics, U.S. House of Representatives, 93rd Congress, Second Session, October 3, 4, and 9, 1974 (Washington, D.C.: Government Printing Office, 1974); A. Warren, "East Africa," in B. W. Hodder and D. R. Harris, eds., *Africa in Transition* (London: Methuen, 1967); and R. A. Pullan, "The Soil Resources of West Africa," in M. F. Thomas and G. W. Whittington, eds., *Environment and Land Use in Africa* (London: Methuen, 1969).

6. Barry Floyd, "Soil Erosion and Deterioration in Eastern Nigeria: A Geographic Appraisal," *Nigerian Geographical Journal*, Vol. 8 (1965). See also A. T. Grove, "Soil Erosion and Population Problems in South-East Nigeria," *Geographical Journal*, Vol. CXVII, pt. 3 (September, 1951); and G. E. K. Ofomata, "Soil Erosion in the Enugu Region of Nigeria," *African Soils*, Vol. IX, No. 2 (May–August, 1964).

7. Donald E. Vermeer, "Population Pressure and Crop Rotational Changes Among the Tiv of Nigeria," *Annals of the Association of American Geographers*, Vol. 60, No. 2 (June, 1970).

8. F. W. Hauck, "Introduction," in Food and Agricultural Organization, *Shifting Cultivation and Soil Conservation in Africa*, Soils Bulletin 24 (Rome: 1974); and Martin Billings, "The African Development Problem," mimeograph (Washington, D.C.: U.S. Agency for International Development, May 24, 1974).

9. U.S. Department of Agriculture, Economic Research Service, *The World Food Situation and Prospects to 1985*, Foreign Agricultural Economic Report No. 98 (Washington, D.C.: December, 1974), p. 15.

10. Quoted in *World Environment Report*, Vol. I, No. 8 (May 12, 1975).

11. R. L. Beuekenkamp, "Amazonia," Parts 1 and 2, *Foreign Agriculture*, September 15 and 22, 1975.

12. Harald Sioli, "Recent Human Activities in the Brazilian Amazon Region and their Ecological Effects," in Betty J. Meggers, Edward S. Ayensu, and W. Donald Duckworth, eds., *Tropical Forest Ecosystems in Africa and South America: A Comparative Review* (Washington, D.C.: Smithsonian Institution Press, 1973), pp. 326–328. See also Frances M. Foland, "A Profile of Amazonia: Its Possibilities for Development," *Journal of Inter-American Studies and World Affairs*, Vol. XIII, No. 1 (January, 1971), p. 76.

13. Harald Sioli, *op. cit.*

14. Frances M. Foland, *op. cit.*, pp. 71–73; R.L. Beuekenkamp, *op. cit.;* and Charles Wagley, "Introduction," in Charles Wagley, ed., *Man in the Amazon* (Gainesville: University Presses of Florida, 1974), p. 9.

15. Brazilian geographers quoted in Betty J. Meggers, *Amazonia: Man And Culture in a Counterfeit Paradise* (Chicago: Aldine-Atherton, 1971), p. 154. William M. Denevan, "Development and the Imminent Demise of the Amazon Rain Forest," *Professional Geographer*, Vol. 25, No. 2 (May, 1973).

16. D. J. Greenland, "Shifting Cultivation vs. Systems for Continuous Production in Arable Lands in the Tropics," Mimeographed, presented to Technical Advisory Committee, Consultative Group on International Agricultural Research, Eighth Meeting, Washington, D.C., July 24–August 2, 1974. Rattan Lal, "Soil Erosion and Shifting Agriculture," in Food and Agriculture Organization, *Shifting Cultivation and Soil Conservation in Africa, op. cit.*, reviews the erosion-reducing techniques being tested at the International Institute for Tropical Agriculture.

17. Jen-Hu Chang, "The Agricultural Potential of the Humid Tropics," *Geographical Review*, Vol. LVIII, No. 3 (July, 1968), p. 360.

18. K. F. S. King, *Agri-Silviculture (The Taungya System)*, Bulletin No. 1 (University of Ibadan, Department of Forestry, 1968); L. Roche, "The Practice of Agri-Silviculture in the Tropics with Special Reference to Nigeria," and E. E. Enabor, "Socio-Economic Aspects of Taungya in Relation to Traditional Shifting Cultivation in Tropical Developing Countries," both in Food and Agriculture Organization, *Shifting Cultivation and Soil Conservation in Africa, op. cit.*

9.   Dual Threat to World Fisheries

1. John Gulland, "The Harvest of the Sea," in William W. Murdoch, ed., *Resources, Pollution and Society*, 2nd ed. (Sunderland, Mass.: Sinauer Associates, 1975).
2. Recent data supplied by Fishery Resources and Environment Division, Food and Agriculture Organization. Historical data on world fish catch from FAO, *Yearbook of Fisheries Statistics*, various issues.
3. See Food and Agriculture Organization, Department of Fisheries, "Review of the Status of Some Heavily Exploited Fish Stocks" (Rome: FAO Fisheries Circular No. 313, 1973).
4. Catch data for northwest Atlantic supplied by International Commission for the Northwest Atlantic Fisheries.
5. Gerald J. Paulik, "Anchovies, Birds and Fishermen in the Peru Current," in William W. Murdoch, ed., *Environment: Resources, Pollution and Society*, 1st ed.(Stamford, Conn.: Sinauer Associates, 1971), p. 157.
6. John Gulland, *op. cit.*, p. 181.
7. Numerous examples of pollution damage or contamination of fisheries are described in Food and Agriculture Organization, *Pollution: An International Problem for Fisheries* (Rome: 1971). Many of the following examples are drawn from this volume.
8. Egyptian and Mexican examples *ibid.* Bangladesh kills noted in *World Environment Report*, Vol. 1, No. 12 (July 7, 1975). On Great Rift Lakes see W. J. Lusigi, "Some Environmental Factors in Food Production in Kenya," prepared for the U.N. World Food Conference, Rome, November 5–16, 1974 (Nairobi: National Environment Secretariat, Office of the President, 1974).
9. See, for example, Gordon Conway and Jeff Romm, *Ecology and Resource Development in Southeast Asia*, report to the Ford Foundation, Office for Southeast Asia (New York: Ford Foundation, August, 1973), p. 72.
10. *Man's Impact on the Global Environment*, Report of the Study of Critical Environmental Problems (Cambridge, Mass.: Massachusetts Institute of Technology, 1970), p. 135.
11. According to Dr. G. M. Woodwell, Marine Biological Laboratory, Woods Hole, Massachusetts. See *New York Times*, February 14, 1975.
12. National Academy of Sciences, *Petroleum in the Marine Environment* (Washington, D.C.: 1975), p. 104. See also M. Blumer, "Oil Contamination and the Living Resources of the Sea," presented to FAO Technical Conference on Marine Pollution and Its Effects on Living Resources and Fishing, Rome, December 9–18, 1970. FIR: MP/70/R–1, September 14, 1970.
13. FAO, *op. cit.*, p. 17.
14. See Michael H. Horn, John M. Teal, and Richard H. Backus, "Petroleum Lumps on the Surface of the Sea," *Science*, Vol. 168, April 10, 1970; James N. Butler, "Pelagic Tar," *Scientific American*, Vol. 232, No. 6 (June, 1975); and National Academy of Sciences, *op. cit.*
15. Robert Citron, "Statement to the Ocean Policy Study of the Committee on Commerce, U.S. Senate," January 30, 1975.
16. *Ibid.*

17. John H. Ryther, "Is the World's Oxygen Supply Threatened?" *Nature*, Vol. 227 (July 25, 1970).
18. Noël Mostert, *Supership* (New York: Alfred A. Knopf, 1974), p. 228.

10.   The Politics of Soil Conservation

1. From John Woodward Thomas, "Employment Creating Public Works Programs: Observations on Political and Social Dimensions," in Edgar O. Edwards, ed., *Employment in Developing Nations* (New York: Columbia University Press, 1974), p. 307.
2. Howard Edward Daugherty, *Man-Induced Ecologic Change in El Salvador*, Ph.D. dissertation, University of California, Los Angeles, Department of Geography, 1969. Unless otherwise noted, the facts on El Salvador that follow are drawn from this study.
3. Organization of American States, *El Salvador Zonificación Agrícola (Fase I)* (Washington, D.C.: OAS, 1974).
4. United Nations Development Program, "Country and Intercountry Programming: Haiti" (New York: U.N.D.P., December 11, 1972); Alan Riding, "Haiti Losing Fight Against Erosion," *New York Times*, June 23, 1974; and Alan Riding, "Saving Haiti," *Saturday Review*, May 17, 1975.
5. World Food Program, "Progress Report on Approved Project: Colombia 498—Reforestation in Ayapel, Department of Cordoba," WFP/IGC: 22/10 Add. 3 (Rome: September, 1972).
6. Hans O. Spielmann, "Probleme der Agrarwirtschaftlichen Entwicklung in Costa Rica," *Die Erde: Zeitschrift der Gesellschaft für Erdkunde*, Vol. 104, No. 1 (1973).
7. See, for example, J. R. V. Prescott, "Overpopulation and Overstocking in the Native Areas of Matabeleland," in R. Mansell Prothero, ed., *People and Land in Africa South of the Sahara: Readings in Social Geography* (New York: Oxford University Press, 1972). Reprinted from *Geographical Journal*, Vol. 127 (1961); and William Allan, *The African Husbandman* (New York: Barnes & Noble, 1965), p. 426.
8. David A. Preston, "The Revolutionary Landscape of Highland Bolivia," *Geographical Journal*, Vol. 135, pt. 1 (March, 1969); and Peter Sartorius and Hans Henle, *Forestry and Economic Development* (New York: Praeger, 1968), p. 9.
9. North Central Farm Management Research Committee, *Conservation Problems and Achievements on Selected Midwestern Farms*, Special Circular 86 (Wooster: Ohio Agricultural Experiment Station, July, 1951), p. 19. Quoted in R. Burnell Held and Marion Clawson, *Soil Conservation in Perspective* (Baltimore: Johns Hopkins University Press, for Resources for the Future, 1965).
10. World Bank, *Land Reform*, World Bank Paper, Rural Development Series (Washington, D.C.: July, 1974), pp. 64–65.
11. Council for Agricultural Science and Technology, *Conservation of the Land, and the Use of Waste Materials for Man's Benefits*, prepared for Committee on Agriculture and Forestry, United States Senate, March 25, 1975 (Washington, D.C.: Government Printing Office, 1975), p. 14.
12. Quoted in D. R. F. Taylor, "Agricultural Change in Kikuyuland," in M. F. Thomas and G. W. Whittington, *Environment and Land Use in Africa* (London: Methuen, 1969), p. 471. See also Paul H. Temple, "Soil and Water Conservation Policies in the Uluguru Mountains, Tanzania," and Len Berry and Janet Townshend, "Soil Conservation Policies in the Semi-Arid Regions of Tanzania, A Historical Perspective," in Anders Rapp, Len Berry, and Paul Temple, eds., *Studies of Soil Erosion and Sedimentation in Tanzania* (University of Dar es Salaam, Bureau of Resource Assessment and Land Use Planning, and University

of Uppsala, Department of Physical Geography, 1972); and Barry Floyd, "Soil Erosion and Deterioration in Eastern Nigeria: A Geographical Appraisal," *Nigerian Geographical Journal*, Vol. 8 (1965), pp. 40–43.

11. *Environmental Stress, Food, and the Human Prospect*

1. United Nations World Food Conference, *Assessment of the World Food Situation, Present and Future.* Rome, November, 5–16, 1974. E/Conf. 65/3.
2. Council for Agricultural Science and Technology, *Conservation of the Land, and the Use of Waste Materials for Man's Benefits*, prepared for Committee on Agriculture and Forestry, United States Senate, March 25, 1975, p. 11; and John Waterbury, "Egypt's Staff of Life," *Common Ground*, Vol. 1, No. 3 (July, 1975).

Additional Sources

1. The Undermining of Food Production Systems

Beasley, R. P. *Erosion and Sediment Pollution Control*. Ames: Iowa State University Press, 1972.

Bennett, Hugh Hammond. "Soil Erosion in Spain." *Geographical Review*, Vol. 50, No. 1 (January, 1960).

Bennett, Hugh Hammond, and Chapline, W. R. "Soil Erosion a National Menace." U.S. Department of Agriculture Circular No. 33. Washington, D.C.: April, 1928.

Brooke, Clarke. "Food Shortages in Tanzania." *Geographical Review*, Vol. 57, No. 3 (July, 1967).

Brown, Leslie H. *Conservation for Survival: Ethiopia's Choice*. Addis Ababa: Haile Sellassie I University, 1973.

Curry-Lindahl, Kai. *Conservation for Survival: An Ecological Strategy*. New York: William Morrow, 1972.

_____. "The Conservation Story in Africa During the 1960s." *Biological Conservation*, Vol. 6, No. 3 (July, 1974).

Dale, Tom, and Carter, Vernon Gill. *Topsoil and Civilization*. Rev. ed. Norman: University of Oklahoma Press, 1974. Originally published 1955.

Dasmann, Raymond F. *Environmental Conservation*. 3rd ed. New York: John Wiley, 1972.

Dasmann, Raymond F.; Milton, John P.; and Freeman, Peter H. *Ecological Principles for Economic Development*. New York: John Wiley, 1973.

Daugherty, Howard E. *Conservación Ambiental Ecológica de El Salvador con Recomendaciones para un Programa de Acción Nacional*. San Salvador: Artes Gráficas Publicitarias, 1973. Also published in English by the Conservation Foundation, Washington, D.C., 1973.

Detwyler, T. R., ed. *Man's Impact on Environment*. New York: McGraw-Hill, 1971.

Dooge, James C. I.; Costin, A. B.; and Finkel, Herman J. *Man's Influence on the Hydrological Cycle*. Irrigation and Drainage Paper 17. Rome: FAO, 1973.

Dorst, Jean. *Before Nature Dies*. Translated by Constance D. Sherman. Boston: Houghton Mifflin, 1970.

Dworkin, Daniel M., ed., *Environment and Development*. SCOPE/UNEP Symposium on Environmental Sciences in Developing Countries, Nairobi, February 11–23, 1974. Indianapolis: SCOPE Misc. Publication, 1974.

Ehrlich, Paul R. "Human Population and Environmental Problems." *Environmental Conservation*, Vol. 1, No. 1 (Spring, 1974).

Ehrlich, Paul R., and Ehrlich, Anne H. *Population, Resources, Environment.* 2nd ed. San Francisco: W. H. Freeman, 1972.

Ehrlich, Paul R.; Ehrlich, Anne H.; and Holdren, John P. *Human Ecology: Problems and Solutions.* San Francisco: W. H. Freeman, 1973.

Farvar, M. Taghi, and Milton, John P., eds. *The Careless Technology: Ecology and International Development.* Garden City, Natural History Press, 1972.

Flannery, Robert D. "The Ecological Undermining of Food Producing Systems in Latin America Caused by Soil Erosion." Unpublished. Guatemala, 1974.

FAO/UNEP. *A World Assessment of Soil Degradation: An International Programme of Soil Conservation.* Report of an Expert Consultation on Soil Degradation, Rome, June, 10–14, 1974.

Graf, H. "Bodenerosion und Schutzmansnahmen im marokkanischen Rifgebirge." *Schweizerische Zeitschrift für Forstwesen,* Vol. 124, No. 11 (November, 1973).

Hart, John, and Socolow, Robert H. *Patient Earth.* New York: Holt, Rinehart and Winston, 1971.

Holdren, John P., and Ehrlich, Paul R., "Human Population and the Global Environment." *American Scientist,* Vol. 62, No. 3 (May–June, 1974).

Hudson, Norman. *Soil Conservation.* Ithaca: Cornell University Press, 1971.

Hughes, J. Donald. *Ecology in Ancient Civilizations.* Albuquerque: University of New Mexico Press, 1975.

Inter-Bureau Mobile Seminar on Ecological Inputs in the Management of Public Works. Manila: National Committee on Man and the Biosphere, UNESCO National Commission of the Philippines, March 24, 1975.

Judson, Sheldon. "Erosion of the Land, or What's Happening to Our Continents?" *American Scientist,* Vol. 56, No. 4 (1968).

Lentnek, Barry; Carmin, Robert L.; and Martinson, Tom L., eds. *Geographic Research in Latin America Benchmark 1970.* Proceedings of the Conference of Latin Americanist Geographers, Vol. 1. Muncie, Ind.: Ball State University, 1971.

Lowdermilk, W. C. "Conquest of the Land Through Seven Thousand Years." U.S. Department of Agriculture, Agriculture Information Bulletin No. 99. Washington, D.C.: August, 1953.

Macias, G. B., and Cervantes, G. R. *La Conservación del Suelo y el Agua en México.* Mexico, D.F.: Instituto Mexicano de Recursos Naturales Renovables, 1966.

Mahler, P.J. "Agricultural Development and the Environment." *Geoforum,* October, 1972.

Marsh, George Perkins. *Man and Nature.* Cambridge: Harvard University Press, 1965. Originally published 1864.

National Academy of Sciences. *Productive Agriculture and a Quality Environment.* Washington, D.C.: 1974.

Olson, Gerald W. "Improving Uses of Soils in Latin America." *Geoderma,* No. 9 (1973).

Organization of American States. *Physical Resource Investigations for Economic Development: A Casebook of OAS Field Experience in Latin America.* Washington, D.C.: 1969.

———. *Scotland District of Barbados: Evaluation of the Problems and Treatment of Erosion and Unstable Ground.* Washington, D.C.: 1971.

———. *Survey of the Natural Resources of the Dominican Republic.* Washington, D.C.: 1969.

Pereira, H. C. *Land Use and Water Resources in Temperate and Tropical Climates.* Cambridge: University Press, 1973.

Polunin, Nicholas. "Thoughts on Some Conceivable Ecodisasters." *Environmental Conservation,* Vol. 1, No. 3 (Autumn, 1974).

———, ed. *The Environmental Future.* New York: Macmillan, 1972.

Proceedings of the Mobile Seminar on Ecological Research Inputs on Soil Resources

Management. Manila: National Committee on Man and the Biosphere, UNESCO National Commission of the Philippines, July, 1974.

Prothero, R. Mansell. *People and Land in Africa South of the Sahara.* New York: Oxford University Press, 1972.

Rapp, Anders. "Soil Erosion and Sedimentation in Tanzania and Lesotho." *Ambio,* Vol. 4, No. 4 (1975).

Ratcliffe, F. *Flying Fox and Drifting Sand.* London: Argus and Robertson, 1948.

Rauschkolb, Roy S. *Land Degradation.* FAO Soils Bulletin 13. Rome: 1971.

República del Perú, Oficina Nacionál de Evaluación de Recursos Naturales, and Organización de los Estados Americanos. *Lineamientos de Política de Conservación de los Recursos Naturales del Perú.* Lima: May, 1974.

Republic of Kenya. *Kenya's National Report to the United Nations on the Human Environment.* Nairobi: April, 1972.

Sasson, Albert. *Développement et Environnement.* Paris: Éditions Mouton, 1974.

Sauer, Carl Ortwin. *Land and Life.* Edited by John Leighly. Berkeley: University of California Press, 1963.

_____. *The Living Landscape.* New York: Basic Books, 1966.

Soemarwoto, Otto. "Environmental Quality: A Challenge in Economic Development." Bandung: Institute of Ecology, Padjadjaran University, August, 1974.

Stallings, J. H. *Soil Conservation.* Englewood Cliffs: Prentice-Hall, 1957.

Stamp, L. Dudley. *Land for Tomorrow: The Underdeveloped World.* Bloomington: Indiana University Press, 1952.

Sternberg, Hilgard O'Reilly. "Man and Environmental Change in South America." In E. J. Fittkau, *et al.,* eds., *Biogeography and Ecology in South America.* Monographiae Biologicae, ed. P. Van Oye, Vol. 18. The Hague: Dr. W. Jungk, 1968.

Thomas, M. F., and Whittington, G. W. *Environment and Land Use in Africa.* London: Methuen, 1969.

Thomas, William L., Jr., ed. *Man's Role in Changing the Face of the Earth.* 2 vols. Chicago: University of Chicago Press, 1956.

UNESCO, National Commission of the Philippines. National Committee on Man and the Biosphere. *A Proposal for Participation in the Man and the Biosphere Programme, Philippines.* Manila: August 1, 1973.

United Nations Environment Program. *Overviews in the Priority Subject Area: Land, Water and Desertification.* Nairobi: February, 1975.

_____. *Review of the Environmental Situation and of Activities Relating to the Environment Programme. Report of the Executive Director.* Governing Council, Third Session, Nairobi, April 17,–May 2, 1975.

U.S. Department of Agriculture. "Mulch Tillage in Modern Farming." Leaflet No. 554. Washington, D.C.: February, 1971.

_____. *Soil,* Yearbook of Agriculture 1957. Washington, D.C.: 1957.

_____. "Soil Erosion: The Work of Uncontrolled Water." Agriculture Information Bulletin 260. Rev. ed. Washington, D.C.: December, 1971.

_____. *Water.* In *Yearbook of Agriculture 1955.* Washington, D.C.: 1955.

Vogt, William. *The Population of Costa Rica and Its Natural Resources.* Washington, D.C.: Pan American Union, 1946.

_____. *The Population of El Salvador and Its Natural Resources.* Washington, D.C.: Pan American Union, 1946.

_____. *The Population of Venezuela and Its Natural Resources.* Washington, D.C.: Pan American Union, 1946.

Vohra, B. B. *A Charter for the Land.* New Delhi: Ministry of Agriculture, September, 1972.

_____. *Land and Water Management Problems in India.* Training Volume No. 8., Training Division, Department of Personnel and Administrative Reforms, Cabinet Secretariat. New Delhi: March, 1975.

2.   *A History of Deforestation*

Allen, Robert. "Woodman, Spare Those Trees." *Development Forum*, April, 1975.
Barney, Daniel R. *The Last Stand: Ralph Nader's Study Group Report on the National Forests*. New York: Grossman, 1974.
Behan, R. W. "Forestry and the End of Innocence." *American Forests*. Vol. 81, No. 5. (May, 1975).
Beresford-Pierse, Henry. *Forests, Food and People*. Basic Study No. 20. Rome: FAO, 1968.
Bormann, F. H. "Temperate Forests—A Human Resource in Jeopardy." Presented to Earthcare Conference, New York, June 5–8, 1975. Sponsored by National Audubon Society and Sierra Club.
Carbonnell, M. M. "La Destruction de la Forêt Tropicale par l'Homme." *Bulletin de la Société Française de Photogrammétrie*. No. 8 (December, 1962).
CENTO Conference on Forestry Development Policy. Ankara, June 8–12, 1970. Ankara: Central Treaty Organization, 1971.
Clawson, Marion. *Forests For Whom and For What?* Baltimore: John Hopkins University Press, for Resources for the Future, 1975.
―――, ed. *Forest Policy for the Future: Conflict, Compromise, Consensus*. Washington, D.C.: Resources for the Future, June, 1974.
Daugherty, Howard E. "The Montecristo Cloud-Forest of El Salvador—A Chance for Protection." *Biological Conservation*. Vol. 5. No. 3 (July, 1973).
Eriksen, Wolfgang. "Waldnutzung und Forstwirtschaft in Argentinen: Ein Bietrag zur Forstgeographie des La Plata Landes." *Geographische Zeitschrift*, Vol. 61, No. 4 (1973).
Eriksen, Wolfgang, and Reisch, Erwin. "Agrarkrise und Aufforstung in Misiones: Rezessions und Expansionserscheinungen in Nordostargentinien." *Die Erde: Zeitschift der Gesellschaft für Erdkunde*, Vol. 105, No. 5, 3/4 (1975).
European Forestry Commission. *Tenth Session of the Working Party on the Management of Mountain Watersheds*. Held in Oslo, August 1–11, 1972. Rome: FAO, 1973.
"l'Exploitation Forestiere: de 1972 à 1975." *Europe Outremer*, No. 527 (December, 1973).
Fanshawe, D. C. "The Conservation of Zambia's Natural Woodlands." *Farming in Zambia*, Vol. 4, No. 3 (April, 1969).
FAO. *Demonstration and Training in Forest, Range and Watershed Management—Philippines*. PH1–65–516. Rome: 1970.
―――. "Environmental Aspects of Natural Resource Management: Forestry." U.N. Conference on the Human Environment, Stockholm, 1972. Agenda Item II (a) ii. FO: HE/72/1. Rome: November 5, 1971.
―――. *Report on a Forestry Project Identification Mission for India*. FAO/SWE/TF 93. Rome: 1973.
―――. *World Symposium on Man-Made Forests and Their Industrial Importance*. Entire issue, *Unasylva*, Vol. 21, No. 5, 86–87 (1967).
FAO/Economic Commission for Africa. *Timber Trends and Prospects in Africa*. Rome: FAO, 1967.
FAO/ Economic Commission for Europe. *European Timber Trends and Prospects: A New Appraisal, 1950–1975*. New York: United Nations, 1964.
FAO/ Economic Commission for Latin America. *Latin American Timber Trends and Prospects*. New York: United Nations, 1963.
FAO/UNDP. *Pre-Investment Study on Forest Industries Development, Ceylon*. Final Report, Vol. I. FAO/SF: 60 CE Y-5. Rome: 1969.
―――. *Present Consumption and Future Requirements of Wood in Tanzania*. FO:SF/TAN 15. Technical Report 3. Rome: 1971.
―――. *Report to the Government of Thailand. Present and Future Forest Policy Goals: A Timber Trends Study 1970–2000*. NOTA 3156. Rome: 1972.

Fifth World Forestry Congress Proceedings. Seattle: 1960.

Freeberne, J. D. M. "The Haiho River Basin Project." *Geography*, Vol. 57, Part 3 (July, 1972).

Gill, Tom. *Land Hunger in Mexico.* Washington, D.C.: Charles Lathrop Pack Foundation, 1951.

Gloriod, G. "La Forêt de l'Ést du Gabon." *Bois et Forêts des Tropiques*, No. 155 (Mai–Juin, 1974).

Gopalakrishnan, T. R., *Forests: 2000 A.D. A Long-Range Perspective for India.* Baroda: Operations Research Group, January, 1975.

Gómez-Pompa, A.; Vásquez-Yanes, C.; and Guevara, S. "The Tropical Rain Forest: A Nonrenewable Resource." *Science*, Vol. 177, No. 4051 (September 1, 1972).

Government of India, Ministry of Agriculture. *Fifth Five-Year Plan, Forestry Sector, 1974–79: Report of the Working Group on Forests.* New Delhi, October, 1972.

————. *Interim Report of the National Commission on Agriculture on Production Forestry—Man-Made Forests.* New Delhi: August, 1972.

Gregerson, Hans M., and Contreras, Arnoldo. *U.S. Investment in the Forest-Based Sector in Latin America: Problems and Potentials.* Baltimore: Johns Hopkins Press, for Resources for the Future, 1975.

Haggett, Peter. "Regional and Local Components in the Distribution of Forested Areas in South East Brazil: A Multivariate Approach." *Geographical Journal*, Vol. 130, Part 3 (September, 1964).

Heseltine, Nigel. "The Ecological Basis of Agriculture in Madagascar." *World Crops*, Vol. 25, No. 1 (January/February, 1973).

Horwitz, Eleanor C. J. *Clearcutting: A View from the Top.* Washington, D.C.: Acropolis Books, 1974.

Irland, Lloyd C. *Is Timber Scarce? The Economics of a Renewable Resource.* Yale University School of Forestry and Environmental Studies, Bulletin No. 83. New Haven: Yale University, 1974.

Jacobs, M. "Luzon's Highest Mountains—Conservation Proposed." *Environmental Conservation*, Vol. I, No. 3 (Fall, 1974).

Jones, Paul H. "World Wood Fibre Outlook." Presented to 56th Annual Meeting of the Woodlands Section, Canadian Pulp and Paper Association. Montreal: 1975.

Kartawinata, Kuswata, and Rubini, Atmawidjaja, eds. *Coordinated Study of Lowland Forests of Indonesia.* Bogor: BIOTROP and IPB, 1974.

Kio, P. R. "Shifting Cultivation and Multiple Use of Forest Land in Nigeria." *Commonwealth Forestry Review*, Vol. 51, No. 2, No. 148 (June, 1972).

Kittredge, Joseph. *Forest Influences: The Effects of Woody Vegetation on Climate, Water and Soil.* New York: McGraw-Hill, 1948.

Kowal, Norman Edward. "Shifting Cultivation, Fire, and Pine Forest in the Cordillera Central, Luzon, Philippines." *Ecological Monographs*, Vol. 36, No. 4 (Autumn, 1966).

Kunkle, Samuel H. "Water—Its Quality Often Depends on the Forester." *Unasylva.* Vol. 26, No. 105 (Summer, 1974).

Lanly, J. P. "Régression de la Forêt Dense en Côte d'Ivoire." *Boits et Forêts des Tropiques*, No. 127 (September–October, 1969).

Lawson, George N. "The Case for Conservation in Ghana." *Biological Conservation.* Vol. 4. (July, 1972).

Lowdermilk, W. C. "Sand Rivers of China." *Soil Conservation.* Vol. III, No. 1 (July, 1937).

Malik, O. P. "Extensive Forestry and Intensive Farming—Slogan of the Day." *Indian Forester*, Vol. 99, No. 11 (November, 1973).

Meijer, William. *Indonesian Forests and Land Use Planning.* Report on NSF-AID Travel Grant, May–August 1973. Lexington: University of Kentucky Bookstore, 1975.

————. "Timber Boom and Land-Use Problems in Indonesia." *Environmental Conservation.* Vol. 1, No. 1 (Spring, 1974).

Messines, Jean. "Forest Rehabilitation and Soil Conservation in China." *Unasylva,* Vol. 12, No. 3 (1958).

Mikesell, Marvin W. "Deforestation in Northern Morocco." *Science,* Vol. 132 (August 19, 1960).

Mueller-Dombois, Dieter. "Planned Utilization of the Lowland Tropical Forests." *Nature and Resources,* Vol. VII, No. 4 (December, 1971).

Noel, Georges. "Bois Tropicaux." *Europe Outremer,* No. 531 (April, 1974).

————. "La Patrimoine Forestier." *El Djeich* (Algiers), No. 131 (April, 1974).

Philippine Forestry and Wood Industries Development. Report of the Presidential Committee on Wood Industries Development. Manila: Bureau of Printing, 1972.

Poore, Duncan. "Saving Tropical Rain Forests." *IUCN Bulletin,* Vol. 5, No. 8 (August, 1974).

Proceedings of the Frontier Forests Officers Conference, December 11–16, 1972. Peshawar: North West Frontier Province, Conservator of Forests, 1973.

"Recommendations of the Forestry Conference, 1973." *Indian Forester,* Vol. 100, No. 2 (February, 1974).

Roche, Laurence. "Major Trends and Issues in Forestry Education in Africa: A View from Ibadan." Bulletin 4, Department of Forest Resources Management, University of Ibadan, Nigeria, 1974.

Sartorius, P. "Ueber die Weltversorgung mit Labholz tropischen Ursprungs." *Schweizerische Zeitschrift für Forstwesen,* Vol. 22, No. 6 (June, 1971).

Schroder, Peter. "Grundlagen, Entwicklung, und Bedeuting der Waldwirtschaft in Tunesien und Algerien: Zugleich ein Beitrag zur Systematik der Weltforstwirtschaft." *Mitteilungen der Bundesforschungsanstalt für Forst und Holzwirtschaft,* Vol. 97 (June, 1974).

Seth, V. K. "Forestry in India as a Lever for Development." *Indian Forester,* Vol. 99, No. 3 (March, 1973).

Singh, G. N. "Inter-Sectoral Linkages of Forest Development." *Indian Forester,* Vol. 99, No. 4 (April, 1973).

Singh, R. V. "Forest Conservation Is Necessary for the Development of Agriculture in Hills." *Indian Forester,* Vol. 100, No. 6 (June, 1974).

Sixth World Forestry Congress Proceedings, Madrid, 1966.

Straub, R. "Die Wiederbewaldung Israels." *Schweitzerische Zeitschrift für Forstwesen,* Vol. 118, No. 4 (1967).

UNESCO, Programme on Man and the Biosphere. *Expert Panel on Project No. 1: Ecological Effects of Increasing Human Activities on Tropical and Subtropical Forest Ecosystems.* Paris, May 16–18, 1972. MAB Report Series No. 3. Paris: September 4, 1972.

————. *Expert Panel on Project 2: Ecological Effects of Different Land Uses and Management Practices on Temperate and Mediterranean Forest Landscapes.* Paris, April 16–19, 1974. MAB Report Series No. 19. Paris: September 4, 1974.

U.S. Department of Agriculture, Forest Service, *The Outlook for Timber in the United States.* Forest Resource Report 20. Washington, D.C.: Government Printing Office, July, 1974.

Wacharakitti, S.; Miller, L. D.; and Tom, C. *Tropical Forest Land Use Evolution.* Environmental Engineering Technical Report No. 1. Fort Collins: Colorado State University, Department of Civil Engineering, August, 1975.

Winters, Robert K. *The Forest and Man.* New York: Vantage Press, 1974.

————. "Forestry Beginnings in India." *Journal of Forest History,* Vol. 19, No. 2 (April, 1975).

Wyatt-Smith, J. "The Malayan Forest Department and Conservation." *Nature Conservation in Malaysia, 1961,* special issue of *Malayan Nature Journal.* ·

3. Two Costly Lessons: The Dust Bowl and the Virgin Lands

Carlson, Avis D. "Dust Blowing." *Harper's Magazine*, July, 1935.

Dobkins, D. A., and Beck, Virgil S. "Stabilizing the Dust Bowl." *Soil Conservation*, Vol. 3, No. 6 (December, 1937).

Drouth: A Report on Drouth in the Great Plains and Southwest. Report of the Special Assistant to the President for Public Works Planning. Washington, D.C.: Government Printing Office, October, 1956.

Hibbs, Ben. "The Dust Bowl Can Be Saved." *Saturday Evening Post*, December 18, 1937.

Jackson, W. A. Douglas. "The Virgin and Idle Lands of Western Siberia and Northern Kazakhstan: A Geographical Appraisal." *Geographical Review*, Vol. 46, No. 1 (January, 1956).

Kraenzel, Carl F. *The Great Plains in Transition*. Norman: University of Oklahoma Press, 1955.

_____. "Great Plains: A Region Basically Vulnerable." In Carl Hodge, ed., *Aridity and Man: The Challenge of the Arid Lands in the United States*. Publication No. 74. Washington, D.C.: American Association for the Advancement of Science, 1963.

McDonald, Angus. "Erosion by Wind and Water in Oklahoma." *Soil Conservation*, Vol. 2, No. 10 (April, 1937).

Newport, Fred C., and Hinde, Robert R. "Farming Level Terraces in the Dust Bowl." *Soil Conservation*, Vol. 4, No. 5 (November, 1938).

Svobida, Lawrence. *An Empire of Dust*. Caldwell, Idaho: Caxton Printers, 1940.

Steinbeck, John. *The Grapes of Wrath*. New York: Bantam, 1970. Originally published 1939.

U.S. Department of Agriculture. *The Great Plains Conservation Program: A Progress Report*. Rev. ed. PA669. Washington, D.C.: September 1969.

_____. *Program for the Great Plains*. Miscellaneous Publication No. 709. Washington, D.C.: Government Printing Office, January, 1956.

Whitfield, C. J. "Sand Dunes in the Great Plains." *Soil Conservation*, Vol. 2, No. 9 (March, 1937).

Whitfield, C. J., and Newport, Fred C. "The Reclamation of a Sand Dune Area." *Soil Conservation*, Vol. 3, No. 7 (January, 1938).

Works Progress Administration. *Social Problems of the Drought Area*. Research Bulletins Series V. Washington, D.C.: Government Printing Office, 1937.

4. Encroaching Deserts

Adeyoju, S. K. *A Pre-Investment Survey of the Northern Arid Region of Nigeria*. University of Ibadan, Department of Forestry, March, 1973.

Adeyoju, S. K., and Enabor, E. E. *A Survey of the Drought Affected Areas of Northern Nigeria*. University of Ibadan, Department of Forestry, November, 1973.

"Algerians Planting Green Belt to Halt Advance of Sahara." *New York Times*, February 16, 1975.

Beltran, Enrique, ed. *Las Zonas Áridas del Centro y Noreste de México*. Mexico, D.F.: Instituto Mexicano del Recursos Naturales Renovables, 1964.

Bhimaya, C. P.; Kaul, R. N.; and Ganguli, G. N. "Sand Dune Rehabilitation in Western Rajasthan." *Fifth World Forestry Congress Proceedings*. Seattle, 1960.

Brown, Leslie H. "The Biology of Pastoral Man as a Factor in Conservation." *Biological Conservation*, Vol. 3, No. 2 (January, 1971).

Catinot, R. "Contribution du Forestier a la Lutte Contre la Désertification en Zones Sèches." *Bois et Forêts des Tropiques*, No. 155 (Mai–Juin, 1974).

CENTO Seminar on Agricultural Aspects of Arid and Semi-Arid Zones, Tehran, September 19–23, 1971. Ankara: Central Treaty Organization, 1972.

Cloudsley-Thompson, J. L. "The Expanding Sahara." *Environmental Conservation,* Vol. 1, No. 1 (Spring, 1974).

Delwaulle, J. C. "Désertification de l'Afrique au Sud du Sahara." *Bois et Forêts des Tropiques,* No. 149 (Mai–Juin, 1973).

DuBois, Victor D. "Drought in West Africa: Temporary Famine or Chronic Deficiency?" *Common Ground,* Vol. 1, No. 3 (July, 1975).

Eckholm, Erik P. "Desertification: A World Problem." *Ambio,* Vol. 4, No. 4 (1975).

Evenari, Michael. "Farmers: Ancient and Modern." *Natural History,* Vol. 183, No. 7 (August/September 1974).

FAO. *The Ecological Management of Arid and Semi-Arid Rangelands in Africa and the Near East: An International Programme.* AGPC:MISC/26. Rome: 1974.

———. *Heathland and Sand Dune Afforestation.* FAO/DEN/TF 123. Rome: 1974.

Government of India, Planning Commission. *Integrated Agricultural Development in Drought Prone Areas.* Report by the Task Force on Integrated Rural Development. New Delhi: June, 1973.

Johnson, Douglas L. *The Nature of Nomadism.* Department of Geography Research Paper No. 118, University of Chicago, 1969.

Kassas, M. "The Nile Ecological System: Towards an International Program." *Biological Conservation,* Vol. 4, No. 1 (October, 1971).

———. "A Brief History of Land Use in Mareotis Region, Egypt." *Minerva Biologica,* Vol. 1, No. 4 (October–December, 1972).

Kowal, J. M., and Adeoye, K. B. "An Assessment of Aridity and the Severity of the 1972 Drought in Northern Nigeria and Neighboring Countries." Samaru Research Bulletin 212. Ahmadu Bello University, Institute for Agricultural Research, Samaru. Zaria: 1974.

Kolars, John. "Locational Aspects of Cultural Ecology: The Case of The Goat in Non-Western Agriculture." *Geographical Review,* Vol. 56, No. 4 (October, 1966).

Le Houérou, H. N. "Deterioration of the Ecological Equilibrium in the Arid Zones of North Africa." Mimeographed. Rome: FAO, 1974.

———. "Peut-on Lutter Contre la Désertisation." Presented to Colloque International Sur la Désertification, Nouakchott, AGPC: MISC/22. Rome: FAO, 1973.

McKell, Cyrus M. "Shrubs—A Neglected Resource of Arid Lands." *Science,* Vol. 187 (March 7, 1975).

MacLeod, Norman. "Can the Sahel Be Saved?" *War on Hunger,* June, 1974.

Mascarenhas, Adolfo C. *Drought and the Optimization of Tanzania's Environmental Potential.* SIES Report No. 6. Stockholm: Secretariat for International Ecology, 1974.

Mishra, M. N.; Prasad, Ram; and Bhan, Suraj. "Arid Zone Agriculture." *World Crops,* Vol. 20, No. 6 (December, 1968).

Morgan, T. W., ed. *Nairobi: City and Region.* Nairobi: Oxford University Press, 1967.

Novikoff, Georges. *The Desertisation of Rangelands and Cereal Cultivated Lands in Pre-Saharan Tunisia. A Statement on Some Possible Methods of Control.* SIES Report No. 3. Stockholm: Secretariat for International Ecology, 1974.

National Academy of Sciences. *Arid Lands of Sub-Saharan Africa.* Progress Report, Fiscal Report, and Appendices. Washington, D.C.: 1975.

Odingo, R. S. *Systems of Agricultural Production in the African Areas of Drought Hazard with Special Reference to the Sahelian Zone of West Africa.* SIES Report No. 5, Stockholm: Secretariat for International Ecology, 1974.

National Academy of Sciences. *More Water for Arid Lands: Promising Technologies and Research Opportunities.* Washington, D.C.: 1974.

Peberdy, J.R. "Rangeland." In W. T. W. Morgan, ed., *East Africa: Its Peoples and Resources.* New York: Oxford University Press, 1969.

Report of the Ad Hoc Panel on the Present Interglacial. ICAS 186-FY75. Washington, D.C.: National Science Foundation, August, 1974.
Schiffers, Heinrich. "Die Dürre im Sudan." *Geographische Rundschau,* August, 1974.
"Sécheresse et Désertification, Deux Freins Majeurs au Développement." *Europe Outremer,* No. 518 (March, 1973).
Sherbrooke, Wade C., and Paylore, Patricia. *World Desertification: Cause and Effect. A Literature Review and Annotated Bibliography.* Arid Lands Resource Information Paper No. 3. Tucson: University of Arizona, Office of Arid Lands Studies, 1973.
UNESCO, Programme on Man and The Biosphere. *Expert Panel on Project 3: Impact of Human Activities and Land Use Practice on Grazing Lands: Savanna, Grassland (from Temperate to Arid Areas), Tundra.* Montpelier, October 2–7, 1972. MAB Report Series No. 6. Paris: March 6, 1973.
_____. *Regional Meeting on Integrated Ecological Research and Training Needs in the Sahelian Region.* MAB Report Series No. 18, draft. Paris: June, 1974.
U.S. Agency for International Development, Bureau for Technical Assistance. *Guidelines for Improving Livestock Production on Range Lands.* Technical Series Paper No. 2. Washington, D.C.: February, 1971.
Waterbury, John. "Land, Man, and Development in Algeria." *Common Ground.* Vol. 1, No. 1 (January, 1975).
Widstrand, Carl Gösta. "The Rationale of Nomad Economy." *Ambio,* Vol. 4, No. 4 (1975).

5. *Refugees from Shangri-La: Deteriorating Mountain Environments*

Ahmad, Ghulan, ed. *Report on Panjkora Valley Forest.* Peshawar: Pakistan Forest Institute, 1968.
Armillus, Redvo. "Land Use in Pre-Columbian America." In L. Dudley Stamp, ed., *A History of Land Use in Arid Regions.* Paris: UNESCO, 1961.
Bailey, Charles R. "The Land, the People, and the University: Research and Action in the Naurar Valley." Presented to Ranikhet Workshop, G. B. Pant University of Agriculture and Technology, September 4–6, 1975. New Delhi: Ford Foundation, 1975.
Bhattarai, S. P. "Die Wälder Nepals, besonders in der Landschaft Terai." *Forstarchiv: Zeitschrift für Wissenschaftlichen und Technischen Fortschritt in der Forstwissenschaft,* Vol. 42, No. 3 (March, 1971).
Blair, I. J. (Agronomist, Pasture and Fodder). *Final Report.* Addis Ababa: Institute of Agricultural Research, August, 1970.
Borries, Oscar V. *Inventario de Problemas Ambientales de Bolivia.* Mimeographed. La Paz: Comisión de Reestudio de Recursos Naturales Renovables, Man and Biosphere Program, June, 1974.
Brown, Leslie H. *Conservation for Survival: Ethiopia's Choice.* Addis Ababa: Haile Sellassie I University, 1973.
Crawford, R. M. M.; Wishart, D.; and Campbell, R. M. "A Numerical Analysis of High-Altitude Scrub Vegetation in Relation to Soil Erosion in the Eastern Cordillera of Peru." *Journal of Ecology,* Vol. 58, No. 1 (March, 1970).
Eckholm, Erik P. "Losing Ground in Shangri-La: Report from Kathmandu." *Saturday Review,* October 4, 1975.
_____. "The Deterioration of Mountain Environments." *Science,* Vol. 189 (September 5, 1975).
Entwicklungsprobleme in Bergregionen: 1. Konferenz des Club of Munich. Schriftenreihe des Alpeninstituts, Heft 3. Munich: Kommissionsverlag Geographische Buchhandlung München, 1975.
FAO/UNDP/ Government of Nepal. *Trishuli Watershed Development Project: Re-*

port on Project Results, Conclusions and Recommendations. Kathmandu: UNDP, no date.

Ford, Thomas R. *Man and Land in Peru.* Gainesville: University of Florida Press, 1962.

Goldstein, Melvyn C. "A Report on Limi Panchayat, Humla District, Karnali Zone." Mimeographed. Cleveland: Case Western Reserve University, Department of Anthropology, April, 1975.

Handleman, Howard. *Struggle in the Andes: Peasant Political Mobilization in Peru.* Austin: University of Texas Press, 1975.

Hartwig, F. "Landschaftswandel und Wirtschaftswandel in der Chilenischen Frontera," *Mitteilungen der Bundesforschungsanstalt für Forst un Holzwirtschaft,* Vol. 61 (May, 1966).

Huffnagel, H. P., ed. *Agriculture in Ethiopia.* Rome: FAO, 1961.

Jan, Abeedullah. *Land Use Survey of Siran and Daur Rivers Watersheds.* Peshawar: Pakistan Forest Institute, 1972.

Jest, Corneille, ed. "Hommes et Milieux Himalayens." Special issue of *Objets et Mondes,* Vol. 14, No. 4 (Winter, 1974).

Joshi, K. L. "The Study of Soil Erosion on West Himalayan Terrain." *Science and Culture* (India), Vol. 38, No. 8 (August, 1972).

Kebede, Yilma. "Chilalo Awraja-Arusi." *Ethiopian Geographical Journal.* Vol. V, No. 1 (June, 1967).

Khan, Faquir Muhammad. *Integrated Resource Survey and Development Potentials of Swat River Watershed.* Peshawar: Pakistan Forest Institute, 1971.

Masrur, Anwar, and Khan, Asghar Ali. *Integrated Resource Survey of Kunhar River Watershed.* Peshawar: Pakistan Forest Institute, 1973.

McIntyre, Loren. "The Lost Empire of the Incas." *National Geographic,* Vol. 144, No. 6 (December, 1973).

Messerschmidt, Donald A. "Gurung Shepherds of Lamjung Himal." *Objets et Mondes,* Vol. 14, No. 4 (Winter, 1974).

Monheim, Felix, "Agrarreform und Kolonisation in Peru und Bolivien." *Geographische Zeitschrift,* Vol. 20 (1968).

Okada, Ferdinand E. *Preliminary Report on Regional Development Areas in Nepal.* Kathmandu: National Planning Commission, July, 1970.

Pereira, H. C., ed. *Hydrological Effects of Changes in Land Use in Some East African Catchment Areas.* Special issue of *East African Agricultural and Forestry Journal,* Vol. XXVII (March, 1962).

————. *Soil Erosion in Ethiopia and Proposals for Remedial Action.* Addis Ababa: Institute of Agricultural Research, March 23, 1968.

Population and Development. Proceedings of the Seminar on Population and Development, Centre for Economic Development and Administration, Tribhuvan University, July 28–30, 1971. Kathmandu: 1971.

Proceedings of the First West Pakistan Watershed Management Conference, November 11–13, 1968. Peshawar: Pakistan Forest Institute, 1968.

Sohai, Rameshwar. "Monogram on Soil and Water Conservation in Uttar Pradesh During the Fifth Five-Year Plan." *Indian Forester,* Vol. 79, No. 1 (January, 1973).

Sellassie, T. G. "Harer Awraja." *Ethiopian Geographical Journal,* Vol. V., No. 1 (June, 1967).

UNESCO, Programme on Man and the Biosphere. *Regional Meeting on Integrated Ecological Research and Training Needs in the Andean Region.* La Paz, June 10–15, 1974. MAB Report Series No. 23, draft. Paris: May, 1975.

Weiss, Dieter, *et al.* "Regional Analysis of Kosi Zone/ Eastern Nepal: Working Method for Regional Planning in Nepal." *Socio-Economic Planning Sciences,* Vol. 7 (1973).

6.   *The Other Energy Crisis: Firewood*

"Beating a Drought, Organically." *Organic Gardening and Farming,* January, 1975.
Dhua, S. P. "Need for Organo-Mineral Fertilizer in Tropical Agriculture." *Ecologist,*
     Vol. 5, No. 5 (June, 1975).
Eckholm, Erik P. *The Other Energy Crisis: Firewood.* Worldwatch Paper No. 1.
     Washington, D.C.: Worldwatch Institute, September, 1975.
_____. "The Firewood Crisis." *Natural History,* Vol. 84, No. 8 (October, 1975).
FAO. *Fuelwood Plantations in India.* Occasional Paper No. 5. Rome: December,
     1958.
_____. *Organic Materials as Fertilizers.* Report of the FAO/ SIDA Expert Consulta-
     tion, Rome, December 2–6, 1974. Soils Bulletin 27. Rome: 1975.
Garg, A. C.; Idnani, M. A.; and Abraham, T. P. *Organic Manures.* Technical Bulletin
     No. 32. New Delhi: Indian Council of Agricultural Research, 1971.
Horvath, Ronald J. "Addis Ababa's Eucalyptus Forest." *Journal of Ethiopian Studies,*
     Vol. VI, No. 1 (January, 1968).
Lamoureux, C. H. "Observations on Conservation in Indonesia." In Kuswata Kar-
     tawinata and Rubini Atmawidjaja, eds., *Papers from a Symposium on Coor-
     dinated Study of Lowland Forests of Indonesia.* Darmaga, Bogor, July 2–5,
     1973. Sponsored by SEAMEO Regional Center for Tropical Biology and
     Bogor Agricultural University. Bogor, Indonesia: 1974.
Sagrhiya, K. P. "The Energy Crisis and Forestry." *Indian Forests,* Vol. 100, No. 9
     (September, 1974).
Swaminathan, M. S. "Organic Manures and Integrated Approaches to Plant Nutri-
     tion." Presented to Consultative Group in International Agricultural Re-
     search, Technical Advisory Committee, Eighth Meeting. Washington,
     D.C., July 24–August 2, 1974.
Thulin, S. "Wood Requirements in the Savanna Region of Nigeria." Savanna Forestry
     Research Station, Nigeria, Tech. Rep. 1. Rome: FAO, 1970.
"World of Forestry." *Unasylva,* Vol. 26, No. 105 (Summer, 1974).

7.   *The Salting and Silting of Irrigation Systems*

Casey, Hugh E. *Salinity Problems in Arid Lands Irrigation: A Literature Review and
     Selected Bibliography.* Prepared by Office of Arid Lands Studies, University
     of Arizona, for Office of Water Resources Research, U.S. Department of the
     Interior, 1972.
Eckholm, Erik P. "Salting the Earth." *Environment,* Vol. 17, No. 7 (October, 1975).
FAO/UNESCO. *Irrigation, Drainage and Salinity: An International Source Book.*
     London: Hutchinson/FAO/UNESCO, 1973.
Framji, K. K., and Mahajan, I. K. *Irrigation and Drainage in the World: A Global
     Review.* 2nd ed. New Delhi: International Commission on Irrigation and
     Drainage, 1969.
Furon, Raymond. *The Problem of Water: A World Study.* Translated by Paul Barnes.
     New York: American Elsevier, 1967.
Green, Donald E. *Land of the Underground Rain.* Austin: University of Texas Press,
     1973.
Holý, Miloš. *Water and the Environment.* Irrigation and Drainage Paper 8. Rome:
     FAO, 1971.
International Commission on Irrigation and Drainage. *Sixth Congress on Irrigation
     and Drainage, Transactions.* New Delhi, 1966.
Michel, Aloys A. *The Kabul, Kunduz, and Helmand Valleys and the National
     Economy of Afghanistan.* Washington, D.C.: National Academy of
     Sciences, 1959.

Mohindra, V. "Reclamation of Saline and Alkali Soils." *Indian Forester,* Vol. 99, No. 3 (March, 1973).

Nace, Raymond. "Man and Water: A Lesson in History." *Bulletin of the Atomic Scientists,* March, 1972.

National Academy of Sciences. *More Water for Arid Lands.* Washington, D.C., 1974.

Nitz, H. J. "Zur Geographie der Künstlichen Bewässerung im Mittleren Nordindien." *Geographische Zeitschrift,* Vol. 4 (December, 1968).

Ravenholt, Albert. "Bali, Microcosm for Third World Agriculture." *Common Ground,* Vol. 1, No. 3 (July, 1975).

Resnick, Sol D., and DeCook, K. J. "Brackish Water as a Factor in Development of the Safford Valley, Arizona, U.S.A." Presented to International Symposium on Brackish Water as a Factor in Development, Ben-Gurion University of the Negev, Beer-Sheva, Israel, January 5–10, 1975.

Robinson, Frank E. *Salinity Management Options for the Colorado River.* California Agricultural Experiment Station, University of California, Davis and El Centro, 1974.

Schroo, H. "Notes on the Reclamation of Salt-Affected Soils in the Indus Plain of West Pakistan." *Netherlands Journal of Agricultural Science,* Vol. 15 (1967).

Skogerboe, Gaylord V. "Improving Water Management to Alleviate Waterlogging and Salinity Problems in Pakistan." Mimeographed. Fort Collins: Agricultural Engineering Department, Colorado State University, February, 1974.

U.S. Department of the Interior. *Report on Water for Energy in the Upper Colorado Basin.* Washington, D.C., July, 1974.

U.S. Salinity Laboratory Staff. *Diagnosis and Improvement of Saline and Alkali Soils.* Agriculture Handbook No. 60. Washington, D.C.: U.S. Department of Agriculture, February, 1954.

Wadleigh, Cecil H. *Wastes in Relation to Agriculture and Forestry.* Miscellaneous Publication No. 1065. Washington, D.C.: U.S. Department of Agriculture, March, 1968.

Waterbury, John. "Egypt's Staff of Life," *Common Ground,* Vol. 1, No. 3 (July, 1975).

Willrich, Ted L., and Smith, George E., eds. *Agricultural Practices and Water Quality.* Ames: Iowa State University Press, 1970.

8. Myth and Reality in the Humid Tropics

Agro-Economic Research on Tropical Soils: Annual Report for 1973. Soil Science Department, North Carolina State University. Raleigh: 1974.

Anderson, Alan. "Farming the Amazon: The Devastation Technique." *Saturday Review,* Vol. 55, No. 40 (October, 1972).

Budowski, Gerardo. "Tropical Savannas, A Sequence of Forest Felling and Repeated Burnings." *Turrialba,* Vol. 6, Nos. 1–2 (January–June, 1956).

de Wilde, John C. *Experiences with Agricultural Development in Tropical Africa.* Vols. I and II. Baltimore: The Johns Hopkins University Press, for the International Bank for Reconstruction and Development, 1967.

Dickinson, J. C. III. "Alternatives to Monoculture in the Humid Tropics of Latin America." *Professional Geographer,* Vol. 24, No. 3 (August, 1972).

Eckholm, Erik P., and Newland, Kathleen. "No Breadbasket in the Jungle." *Development Forum,* Vol. 3, No. 7 (October, 1975).

Farming Systems Program: 1973 Report. Ibadan: IITA, 1974.

Farnworth, Edward G., and Golley, Frank B., eds. *Fragile Ecosystems: Evaluation of Research and Applications in the Neotropics.* A Report of the Institute of Ecology, June, 1973. New York: Springer-Verlag, 1974.

Fournier, F. "Research on Soil Erosion and Soil Conservation in Africa," *African Soils,*
  Vol. XII, No. 1 (January–April, 1967).
Geertz, Clifford. *Agricultural Involution: The Processes of Ecological Change in In-
  donesia.* Berkeley and Los Angeles: University of California Press, 1963.
Goodland, Robert J. A., and Irwin, Howard S. *Amazon Jungle: Green Hell to Red
  Desert?* New York: American Elsevier, 1975.
Gourou, Pierre. *The Tropical World.* Translated by E. D. Laborde. 3rd ed. New York:
  John Wiley, 1961.
Hailey, Lord. *An African Survey: A Study of Problems Arising in Africa South of the
  Sahara.* London: Oxford University Press, 1938.
Harris, D. R. "Tropical Vegetation: An Outline and Some Misconceptions," *Geogra-
  phy,* Vol. 59, pt. 3 (July, 1974).
Jahoda, John C., and O'Hearn, Donna L. "The Reluctant Amazon Basin." *Environ-
  ment,* Vol. 17, No. 7 (October, 1975).
Janzen, Daniel H. "Tropical Agroecosystems." *Science,* Vol. 182 (December 21,
  1973).
Jurion, F., and Henry, J. *Can Primitive Farming Be Modernized?* Brussels: INEAC,
  1969.
Lal, R. "Soil Management Systems and Erosion Control." Presented to Workshop on
  Soil Conservation and Management in the Humid Tropics, Ibadan, IITA,
  June 30–July 4, 1975.
Miracle, Marvin. *Agriculture in the Congo Basin.* Madison: University of Wisconsin
  Press, 1967.
Moraes, Vicente H. F. "Case Study from Brazil." In *Environmental Accomplishments
  to Date.* Columbus, Ohio: Battelle Memorial Institute, 1975.
National Academy of Sciences. *African Agricultural Research Capabilities.* Washing-
  ton, D.C., 1974.
————. *Soils of the Humid Tropics.* Washington, D.C., 1972.
Nelson, Michael. *The Development of Tropical Lands: Policy Issues in Latin America.*
  Baltimore: The Johns Hopkins University Press, for Resources for the Fu-
  ture, 1973.
Phillips, John. *Agriculture and Ecology in Africa.* London: Faber & Faber, 1959.
————. *The Development of Agriculture and Forestry in the Tropics: Patterns, Prob-
  lems, and Promise.* London: Faber & Faber, 1961.
Poleman, Thomas T. *The Papaloapan Project: Agricultural Development in the Mexi-
  can Tropics.* Stanford: Stanford University Press, 1964.
Poore, Duncan. *Ecological Guidelines for Development in Tropical Forest Areas of
  South East Asia.* Occasional Paper No. 10. Morges, Switzerland: Interna-
  tional Union for Conservation of Nature and Natural Resources, 1974.
Roose, E. J. "Quelques Techniques Autiérosives Appropriées aux Régions Tropicales."
  Presented to Workshop on Soil Conservation and Management in the
  Humid Tropics, Ibadan, IITA, June 30–July 4, 1975.
Rosenbaum, H. Jon, and Tyler, William G. "Policy-Making for the Brazilian Ama-
  zon." *Journal of Inter-American Studies and World Affairs,* Vol. 13, Nos. 3
  and 4 (July–October, 1971).
Sánchez, Pedro A., ed. *A Review of Soils Research in Tropical Latin America.* Techni-
  cal Bulletin No. 219. Raleigh: North Carolina Agricultural Experiment
  Station, North Carolina State University, July, 1973.
Sánchez, Pedro A., and Buol, S. W. "Soils of the Tropics and the World Food Crisis."
  Science, Vol. 188 (May 9, 1975).
Spencer, J. E. *Shifting Cultivation in Southeastern Asia.* Berkeley and Los Angeles:
  University of California Press, 1966.
Steel, Robert W. "Population Increase and Food Production in Tropical Africa."
  Special issue of *African Affairs,* Spring, 1965.
Sternberg, Hilgard O'Rielly. *The Amazon River of Brazil.* Wiesbaden: Franz Steiner
  Verlag, 1975.

————. "Land and Man in the Tropics." *Proceedings of the Academy of Political Science,* Vol. 27, No. 4 (1964).

Tosi, Joseph A., Jr., and Voertman, Robert F. "Some Environmental Factors in the Economic Development of the Tropics." *Economic Geography,* Vol. 40, No. 3 (July, 1964).

The Use of Ecological Guidelines for Development in the American Humid Tropics. Proceedings of International Meeting held at Caracas, Venezuela, February 20–22, 1974, No. 31. Morges, Switzerland: International Union for Conservation of Nature and Natural Resources, 1975.

The Use of Ecological Guidelines for Development in Tropical Forest Areas of South East Asia. Papers and Proceedings of the Regional Meeting held at Bandung, Indonesia, May 29–June 1, 1974, No. 32. Morges, Switzerland: International Union for Conservation of Nature and Natural Resources, 1975.

Watters, R. F. *Shifting Cultivation in Latin America.* Forestry Development Paper No. 17. Rome: FAO, 1971.

Webster, C. C., and Wilson, P. N. *Agriculture in the Tropics.* London: Longmans, 1966.

Wycherly, P. R. *Conservation in Malaysia.* Supplementary Paper No. 22. Morges, Switzerland: International Union for Conservation of Nature and Natural Resources, 1969.

9. *Dual Threat to World Fisheries*

Boesch, Donald F.; Hershner, Carl H.; and Milgram, Jerome H. *Oil Spills and the Marine Environment.* Cambridge, Mass.; Ballinger, 1974.

Blumer, M. "Oil Contamination and the Living Resources of the Sea." Presented to FAO Technical Conference on Marine Pollution and Its Effects on Living Resources and Fishing, Rome, December 9–18, 1970. FIR: MP/70/R-1, September 14, 1970.

Christy, Francis T., Jr. *Alternative Arrangements for Marine Fisheries: An Overview.* RFF/PISFA Paper 1. Washington, D.C., Resources for the Future, May, 1973.

FAO. *Pollution: An International Problem for Fisheries.* Rome, 1971.

FAO, Department of Fisheries. "Review of the Status of Some Heavily Exploited Fish Stocks." Fisheries Circular No. 313. Rome: 1973.

Freeman, Peter H. *Coastal Zone Pollution by Oil and Other Contaminants: Guidelines for Policy, Assessment and Monitoring in Tropical Regions.* Based upon a case study in Indonesia in 1973. Washington, D.C.: Office of Environmental Programs, Smithsonian Institution, 1974.

Ginsburg, Norton; Holt, Sidney; and Murdoch, William W., eds. *The Mediterranean Marine Environment and the Development of the Region.* Pacem en Maribus III. Published for the International Ocean Institute by the Royal University of Malta Press, 1974.

Hann, Roy W., Jr. *Follow-up Field Survey of the Oil Pollution from the Tanker Metula.*" Report to the U.S. Coast Guard Research and Development Program. College Station, Tex.: Texas A & M University, Civil Engineering Department, January 6–25, 1975.

————. *Oil Pollution from the Tanker "Metula."* Report to the U.S. Coast Guard Research and Development Program. College Station, Texas: Texas A & M University, Civil Engineering Department, August–September 1974.

Hargrove, John Lawrence, ed. *Who Protects the Ocean? Environment and the Development of the Law of the Sea.* St. Paul, Minn.: West Publishing Company, for the American Society of International Law, 1975.

Idyll, C. P. "The Anchovy Crisis." *Scientific American,* 228, No. 6 (June, 1973).

Jarrin, Edgardo Mercado. "Utilizing Sea Resources for Human and Social Welfare." *Pacific Community*, Vol. 3, No. 2 (January 1, 1972).

Knight, H. Gary, ed. *The Future of International Fisheries Management*. St. Paul, Minn.: West Publishing Company, for the American Society of International Law, 1975.

Law of the Sea: Caracas and Beyond. Proceedings of 9th Annual Conference, Law of the Sea Institute, Kingston, Rhode Island, January 6–9, 1975. Cambridge, Mass.: Ballinger, 1975.

Livingston, Dennis. "Oil on the Seas: Two Cheers for a New Treaty." *Environment*, Vol. 16, No. 7 (September, 1974).

Longhurst, Alan, *et al.* "The Instability of Ocean Populations." *New Scientist*, Vol. 54, No. 798 (June 1, 1972).

Luna, Julio. "Fishery Potential of Latin America: Prospects and Tasks Ahead." Washington, D.C.: Inter-American Development Bank, September, 1974.

MacIntyre, Ferrin, and Holmes, R. W. "Ocean Pollution." In William W. Murdoch, ed., *Environment: Resources, Pollution and Society*. 1st ed. Stamford, Conn.: Sinauer, 1971.

Montague, Peter, and Montague, Katherine. "Mercury: How Much Are We Eating?" *Saturday Review*, February 6, 1971.

Ocean Resources and the Ocean Environment. Proceedings of the Caracas Seminars, July 9, 11, 15, 1974. Sierra Club Special Publication, International Series No. 3. New York, 1974.

Odum, W. E. "Insidious Alteration of the Estuarine Environment." *Transactions of the American Fisheries Society*, Vol. 99, No. 4 (1970).

Perspectives on Ocean Policy. Conference on Conflict and Order in Ocean Relations, October 21–24, 1974, Airlie, Virginia. Prepared for National Science Foundation by Ocean Policy Project, The Johns Hopkins University, Washington, D.C. Washington, D.C.: Government Printing Office, 1975.

Pillay, T. V. R. "The Role of Aquaculture in Fishery Development and Management." FAO Technical Conference on Fishery Management and Development, Vancouver, February 13–23, 1973.

Stein, Robert E., ed. *Critical Environmental Issues on the Law of the Sea*. London and Washington, D.C.: International Institute for Environment and Development, 1975.

10. *The Politics of Soil Conservation*

Christy, Lawrence C. *Legislative Principles of Soil Conservation*. Soils Bulletin 15. Rome: FAO, 1971.

Council for Agricultural Science and Technology. *Conservation of the Land, and the Use of Waste Materials for Man's Benefits*. Prepared for Committee on Agriculture and Forestry, United States Senate, March 25, 1975. Washington, D.C.: Government Printing Office, 1975, p. 14.

Crowder, Michael. *West Africa under Colonial Rule*. London: Hutchinson, 1968.

Crowe, Beryl L. "The Tragedy of the Commons Revisited." *Science*, Vol. 166 (November 28, 1969).

Dorner, Peter. *Land Reform and Economic Development*. Baltimore: Penguin, 1972.

Fernea, Robert A. "Land Reform and Ecology in Postrevolutionary Iraq." *Economic Development and Cultural Change*, Vol. 17, No. 3 (April, 1969).

FAO. *Land Reform, Land Settlement and Cooperatives*. 1974, No. 1/2. Rome: 1975.

Grant, James P. *Growth from Below: A People-Oriented Development Strategy*. Development Paper 16. Washington, D.C.: Overseas Development Council, December, 1973.

Hardin, Garrett. "The Tragedy of the Commons." *Science*, Vol. 162 (December 13, 1968).

Leopold, Aldo. "The Land Ethic." In Aldo Leopold, *A Sand Country Almanac*. New York: Sierra Club/Ballantine, 1970. Originally published 1949.

Morgan, Robert. *Governing Soil Conservation: Thirty Years of the New Decentralization*. Baltimore: The Johns Hopkins University Press, for Resources for the Future, 1966.

Nagadevara, V. S.; Heady, E. O.; and Nicol, J. *Implications of Application of Soil Conservancy and Environmental Regulations in Iowa Within a National Framework*. CARD Report No. 57. Ames: Iowa State University, June, 1975.

Orleans, Leo A. "China's Environomics: Backing into Ecological Leadership." In *China: A Reassessment of the Economy*. A compendium of papers submitted to the Joint Economic Committee, Congress of the United States. Washington, D.C.: Government Printing Office, July 10, 1975.

Owens, Edgar, and Shaw, Robert d'A. *Development Reconsidered*. Lexington, Mass.: Lexington Books, 1972.

Segal, Aaron. "Haiti." In Aaron Lee Segal, ed., *Population Policies in the Caribbean*. Lexington, Mass.: Lexington Books, 1975.

Thiesenhusen, William C. "Colonization: Alternative or Supplement to Agrarian Reform." In Peter Dorner, ed., *Land Reform in Latin America: Issues and Cases*. Madison: Land Economics for the Land Tenure Center, University of Wisconsin, 1971.

Wallman, Sandra. *Take Out Hunger: Two Case Studies of Rural Development in Basutoland*. London: University of London, Athlone Press, 1969.

11. *Environmental Stress, Food, and the Human Prospect*

Berg, Alan. "The Trouble with Triage." *New York Times Magazine*, June 15, 1975.

Brown, Lester R. *In the Human Interest: A Strategy to Stabilize World Population*. New York: W. W. Norton, 1974.

———. *World Without Borders*. New York: Vintage, 1972.

Brown, Lester R., and Eckholm, Erik P. *By Bread Alone*. New York: Praeger, 1974.

Buringh, P.; van Heemst, H. D. J.; and Staring, G. J. *Computation of the Absolute Maximum Food Production of the World*. Wageningen, Netherlands: Agricultural University, Department of Tropical Soil Science, January, 1975.

Falk, Richard A. *This Endangered Planet: Prospects and Proposals for Human Survival*. New York: Random House, 1971.

Hardin, Garrett. "Carrying Capacity as an Ethical Concept." In George Lucas, ed., *Lifeboat Ethics*. New York: Harper & Row, 1976.

———. "Living on a Lifeboat." *Bioscience*, Vol. 24, No. 10 (October, 1974).

Heilbroner, Robert. *An Inquiry into the Human Prospect*. New York: W. W. Norton, 1974.

Kocher, James E. *Rural Development, Income Distribution, and Fertility Decline*. New York: Population Council, 1973.

McNamara, Robert S. *One Hundred Countries, Two Billion People: The Dimensions of Development*. New York: Praeger, 1973.

Meadows, Donella H.; Meadows, Dennis L.; Randers, Jørgen; and Behrens, William W. III. *The Limits to Growth*. New York: Universe Books, 1972.

Mesarovic, Mihajlo, and Pestel, Eduard. *Mankind at the Turning Point: The Second Report of the Club of Rome*. New York: Reader's Digest Press, 1974.

National Academy of Sciences. *Agricultural Production Efficiency*. Washington, D.C., 1975.

———. *Population and Food: Crucial Issues*. Washington, D.C., 1975.

Rich, William. *Smaller Families Through Social and Economic Progress.* Monograph
        No. 7. Washington, D.C.: Overseas Development Council, 1973.
Schertz, Lyle P. "World Food: Prices and the Poor." *Foreign Affairs,* Vol. 52, No. 3
        (April, 1974).
Schumacher, E. F. *Small Is Beautiful: Economics as If People Mattered.* New York:
        Harper & Row, 1973.
Science, special issue on food, Vol. 188, No. 4188, May 9, 1975.
Strong, Maurice F. "The Environment and the New Internationalism," 1973 Fairfield
        Osborn Memorial Lecture. Washington, D.C.: The Conservation Founda-
        tion, 1973.
U.S. Department of Agriculture, Economic Research Service. *The World Food Situa-
        tion and Prospects to 1985.* Foreign Agricultural Economic Report No. 98.
        Washington, D.C., December, 1974.
Ward, Barbara. *The Home of Man.* New York: W. W. Norton, 1976.
Ward, Barbara, and Dubos, René. *Only One Earth: The Care and Maintenance of a
        Small Planet.* New York: W. W. Norton, 1972.
World Bank. *Environment and Development.* Washington, D.C., June, 1975.
_____. *The Assault on World Poverty: Problems of Rural Development, Education,
        and Health.* Baltimore: The Johns Hopkins University Press, for the World
        Bank, 1975.

Index